D1785659

INVENTIONS AND DISCOVERIES

A TRIP THROUGH HISTORY

PHILL JONES

ISBN-10: 149038460X
EAN-13: 9781490384603

To my parents, who encouraged my fascination with science.

Table of Contents

About the Author

Phill Jones earned a PhD in physiology and pharmacology from the University of California at San Diego and a JD from the University of Kentucky College of Law. He worked ten years as a patent attorney, specializing in biological, chemical, and medical inventions. He has written more than 350 articles in the areas of science, medicine, history, law and business. His articles related to inventions and discoveries have appeared in *History Magazine*, *Nature Biotechnology*, *Modern Drug Discovery*, *The Scientist*, *Journal of BioLaw & Business*, *Information Systems for Biotechnology News Report*, *Today's Science*, *PharmaTechnology Magazine*, *Columbia: The Magazine of Northwest History*, and *Regulatory Affairs Focus*. For the education market, Phill wrote seven science books and two law books for Chelsea House. His book, *Forensic Science for Writers* (CreateSpace, 2012), was based on his online course. The book won a silver medal in the writing/publishing division of the 2013 Independent Publisher Book Awards contest. He recently published *Criminal Investigators, Villains and Tricksters: A Trip Through History* (2013), which is a companion volume to this book.

Introduction

Some discoveries occur by accident. Charles Darwin discovered principles of evolution as he studied animals, plants and fossils during his five-year voyage around the world. Other discoveries are intentionally achieved. For more than five years, Howard Carter searched for Tutankhamun's tomb, supervising a team of workers who cleared a desert surface down to bedrock. Sometimes, a person applies a discovery of a natural phenomenon to create an invention. Take quartz clocks, for example. In 1880, Pierre Curie discovered that an electrical current causes particular crystals to expand and contract. About 50 years later, Warren Marrison invented the first quartz clock that used the regular vibrations of a quartz crystal in an electrical circuit. Often, a necessity compels inventors. Fulfilling needs drove the creation of many inventions from a treatment for diabetes to the creation of plastics.

In the following pages, you'll explore inventions and discoveries that shaped today's world. You'll discover several strange events, such as the Boston molasses flood and a drifting ghost ship named the *Mary Celeste*. And you'll discover strange ideas, such as the miasma theory, an incorrect belief about the spread of disease, which launched a revolution in public health improvement.

With the exception of "The Smith Butchering Machine," the articles in this book (in revised form) were published by *History Magazine*. Entitled "Revolution on a Dare," the Smith butchering machine article appeared (again, in revised form) in *Columbia: The Magazine of Northwest History*. The last section of the book offers suggestions for further reading about topics covered in the longer articles.

I. Science

The Discovery of Tutankhamun's Tomb

On November 4, 1922, a young, restless water boy made history at an excavation in Egypt's Valley of the Kings. Imitating his elders, he dug into the hot sand. Soon, he hit a hard surface – a stone step. The boy ran across the site and told Howard Carter about his finding. By the end of the day, workers uncovered a stone stairway that led to an ancient tomb.

Carter sent a telegram to England.

"At last have made wonderful discovery in Valley," he told his benefactor, Lord Carnarvon, "a magnificent tomb with seals intact; re-covered same for your arrival; congratulations."

Before long, Carter learned that the tomb held one of the greatest archeological discoveries of the time: the final resting place of King Tutankhamun.

The Boy King

Amenhotep III, who might have been Tutankhamun's father or grandfather, became pharaoh at the age of 12 and ruled for 38 years. During this golden age of the 18th Dynasty, the Egyptian empire accumulated immense wealth, and excelled in literary and artistic achievements.

As a ruler, the pharaoh's son, Amenhotep IV, failed to measure up to his father. Amenhotep IV abandoned the god Amun and the pantheon of lesser Egyptian gods. Turning from the traditional gods, Amenhotep IV advocated devotion of the Aten, the solar disk. The new pharaoh changed his name to Akhenaten, "servant of the Aten," and he built a new capital city christened Akhetaten, "horizon of the sun disk," which is modern Tel el-Amarna. In doing so, the eccentric king removed power from the administrative headquarters in the city of Memphis and religious center in the city of Thebes.

Smenkhkare succeeded Akhenaten. He ruled for a brief time and quietly left.

Around 1343 BC, a boy of about ten years, Tutankhaten, ascended to the throne and assumed control of the Egyptian superpower. Although his name meant "living image of the Aten," the boy king restored the old ways by reinstating the traditional pantheon of gods and reopening their temples. Tutankhaten also reestablished Memphis and Thebes as seats of power. To honor Amun, Tutankhaten and Ankhesenpaaten, his chief queen, changed their names to Tutankhamun and Ankhsenamun.

Tutankhamun's relatively peaceful nine-year reign ended with his untimely death. Many conspiracy theories evolved to explain how the boy king died. Despite a lack of evidence for murder, two of the pharaoh's advisors have figured in assassination theories: Aye, who might have been Ankhsenamun's grandfather, and General Horemheb, the commander-in-chief of the army.

The burial ritual of Tutankhamun reflected the Egyptian belief that his ba and ka, the two parts of the king's personality, separated from the body. The ka, an individual's life force, needs food, drink, clothing and other earthly requirements to function in the afterlife. The ba, the soul or personality, could leave the tomb and travel around the earth during daylight. A correctly mummified body and properly-executed funerary rites would enable the ba and ka to be reunited, allowing the deceased to become an akh, an eternal spirit.

An elaborate embalming process preserved the body so that the ba could return to the mummy at night and ensure its continued life. Embalmers removed the lungs, liver, stomach, and intestines, dried the organs in salts, anointed them with oils, wrapped the organs, and placed them in solid gold miniature coffins. Using a long metal hook, the embalmers removed the brain through the nostrils.

They dried the corpse's flesh with natron, a mixture of salt and baking soda. After applying resins to soften the skin, embalmers wrapped Tutankhamun's limbs in linen bandages, while priests chanted spells and placed amulets and jewelry between the layers of cloth. They covered the body's fingers and toes with golden cylinders and placed a golden funerary mask on his face.

Aye, Tutankhamun's successor, prepared a small set of rooms in an underground tomb near the floor of the Valley of the Kings, located on the west bank of the Nile and across from Thebes. Soon after the burial, thieves invaded the tomb, but were caught. Officials resealed the vault. In time, workers constructed a tomb for another pharaoh nearby. Their huts obscured Tutankhamun's burial place. Over the years, floods erased any surface evidence of the young king's tomb.

Howard Carter Draws Near His Destiny

Carter was born on May 9, 1874 in London. As a boy, he suffered from ill health and lived with his aunts in rural Swaffham. Carter's parents, who believed their son too delicate to attend a private school, arranged for home education. Carter's father was an artist who worked for the *Illustrated London News* and also specialized in animal paintings. He taught Howard drawing, and found that his son had an aptitude for it.

Howard Carter's interest in Egyptian antiquities and his artistic talent brought him to Egypt in 1891. London's Egypt Exploration Fund hired Carter to help P.E. Newberry record drawings and inscriptions of tombs at Beni Hassan and el-Bersha. During the following decade, Carter gained archaeological experience at the excavation of el-Amarna with Flinders Petrie and as a member of an expedition at the temple of Hatshepsut at Deir el-Bahri.

In 1899, Carter accepted the position of Inspector General of Monuments for Upper Egypt, and controlled archeological work in the Nile Valley. A fight between drunken French tourists and Egyptian guards ended Carter's Antiquities Service career. Carter refused to apologize for his guards. He insisted that his men had only defended themselves; it was the tourists who should apologize. Following a demotion to a minor post, Carter resigned from the Antiquities Service in 1905.

At the same time, George Edward Stanhope Molyneux Herbert, Lord Carnarvon, toured Egypt to recover from a devastating automobile accident. In 1908, Carnarvon decided to finance an archeological exploration.

The government required that such work must be supervised by an experienced archeologist. Carter, who eked out a living as a commercial artist and tour guide, happened to be available.

At first, Carter and Carnarvon focused on Thebes. In 1912, they moved their operation to the Delta with modest results.

The Egyptian government – under British occupation – granted qualified archeologists permission to excavate on ancient sites. The Antiquities Department allowed them to take out of the country half the antiquities that they found. The government excluded the Valley of the Kings from this rule, but did grant one license to explore the area.

Since 1902, Theodore M. Davis, a wealthy American, had secured the license to dig in the Valley of the Kings. In 1906, Davis' archaeologists uncovered a blue glaze cup bearing the cartouche of Tutankhamun. During the next year, they found a rock-cut chamber that held numerous objects with Tutankhamun's name. Davis assumed that he had discovered the tomb of Tutankhamun.

Carnarvon had acquired one of the most valuable private collections of Egyptian antiquities by 1914. Nevertheless, Carter convinced Carnarvon to obtain the concession to explore the Valley of the Kings when Davis, who believed that he had uncovered all major finds, relinquished his license. Carter had been gathering scraps of information on Tutankhamun and thought that the pharaoh's tomb remained hidden in the Valley of the Kings. Although they acquired the license, the outbreak of the First World War forced a postponement of their work.

In 1917, Carter and Carnarvon began their exploration of the Valley of the Kings. Carter decided that the only way to search for Tutankhamun's tomb would be to ignore earlier excavations. He focused on a two and one-half acre triangular plot of land defined by the tombs of Ramesses II, Merenptah, and Ramesses VI. For the first time in the history of Egyptian archeology, he would clear the surface down to the bedrock. To ensure that the work would be systematic, Carter devised a grid system based on the step-by-step artillery barrages of the war. Carter's plan required the transport of hundreds of thousands of cubic meters of sand, rock chips and boulders, labor performed by men and young boys with picks, hoes and small baskets.

For five years, the work yielded little. In the summer of 1922, Carnarvon told Carter that he would no longer fund the expedition. Carter persuaded his benefactor to persist for one more season.

Grave Discovery

On November 1, Carter continued his search in the Valley of the Kings. After a water boy discovered a stone step on November 4, Carter and his workers spent the afternoon uncovering 12 steps of a rock-cut stairway that descended at a 45 degree angle into a small hillock below the entrance to the tomb of Ramesses VI. At the level of the 12th step, Carter found the upper portion of a door constructed of large stones that had been plastered. The doorway's surface bore the Royal Necropolis seal: Anubis over nine foes. Carter could not find a royal name, but he did notice that a corner had been resealed, indicating that robbers had broken into the tomb during ancient times and that something valuable remained.

Carter made a small peephole, inserted an electric light and looked inside. On the other side of the door, he saw a passage filled from floor to ceiling with stones and rubble, a sign that care had been taken to protect the tomb. He ordered his workers to refill the stairway for protection and sent a telegram to Carnarvon in England.

Three weeks later, Lord Carnarvon arrived with his daughter, Lady Evelyn Herbert. After clearing all 16 steps of the stairway, Carter found a seal impression of Tutankhamun on the lower part of the doorway. When the workers cleared the rubble from the corridor, they found a second plastered doorway, which also appeared to have been broken and resealed in antiquity.

On November 26, Carter used his hands to dig a small breach in the second doorway. He inserted an iron rod into the opening and found empty space on the other side. He then lit a candle to check for noxious gases. In his book, *The Tomb of Tut Ankh Amen* (1923), Carter recorded his impressions of the moment:

> At first I could see nothing, the hot air escaping from the chamber causing the candle flame to flicker, but presently, as my eyes grew

accustomed to the light, details of the room within emerged slowly from the mist, strange animals, statues and gold—everywhere the glint of gold. For the moment – an eternity it must have seemed to the others standing by – I was struck dumb with amazement, and when Lord Carnarvon, unable to stand the suspense any longer, inquired anxiously, "Can you see anything?" it was all I could do to get out the words, "Yes, wonderful things."

Carter and Carnarvon entered the room that they would name the Antechamber. Here, they found three large gilt couches with sides carved in the form of animals, and two life-sized figures of a king in black that faced each other like sentinels dressed in gold kilts and gold sandals, and armed with mace and staff. The room also held painted and inlaid caskets, alabaster vases, black shrines, carved chairs, beds, a golden inlaid throne, and a heap of overturned chariots that shimmered with gold and inlay. On the floor, Carter found a large bouquet of flowers with preserved petals and leaves.

Despite the hundreds of treasures it held, the Antechamber measured only 12 by 26 feet with a ceiling seven and one-half feet high. It did not contain a mummy.

Through a plunderer's hole in one wall, they found a ransacked room that they named the Annex. The cluttered chamber held oils, wine, food, carved thrones, an ivory-covered, carved chest, detailed alabaster figures of animals and a boat, game tables, vases and other everyday items that the pharaoh could take with him to the afterlife.

Between the Antechamber's two sentinel statues they found another sealed doorway. Did it lead to the Burial Chamber? Carter and his colleagues secured the site, mounted their donkeys and returned home, silent and subdued.

After notifying the Antiquities Service, Carter assembled an international team of experts to examine the tomb and preserve its contents in drawings and photographs. They offered the first official press viewing of Tutankhamun's tomb on December 22. News of the richest collection of ancient Egyptian treasure sparked a frenzy in the media. Although it had been hastily ransacked, the tomb remained almost intact. For the first time, archeologists could study all of a pharaoh's funerary equipment that offered insights into an ancient culture.

Triumph Begins to Unravel

Tens of thousands of visitors rushed to the Valley of the Kings and interfered with the study of the site. Carter became frustrated and began to turn everyone away from the tomb, including those who had official government permission.

By the end of February, the contents of the Antechamber had been carefully removed for examination. Carter made a hole in the doorway between the sentinels and inserted an electric torch.

"An astonishing sight its light revealed," Carter wrote, "for there, within a yard of the doorway, stretching as far as one could see and blocking the entrance to the chamber, stood what to all appearance was a solid wall of gold."

They removed stones from the doorway, revealing the side of a nine-foot tall, gilt shrine. Within this shrine, they found a second shrine, built to cover a sarcophagus. They had entered the Burial Chamber.

The room contained objects that the king would need during his journey through the underworld: seven oars to ferry himself across the waters of the underworld, lamps of translucent calcite, a silver trumpet, and jars of perfume and unguents. The walls of the chamber were decorated with brightly painted scenes and inscriptions.

Further examination revealed the entrance to yet another chamber: the Treasury. A figure of the jackal god Anubis guarded the entrance to this chamber, which contained a monument, the central portion of which consisted of a large shrine-shaped chest overlaid with gold. A statue of a goddess guarded the shrine on each of its four sides. The chest held jars of preserved organs. The Treasury also held numerous black shrines, chests, and caskets of ivory and wood.

In April, Lord Carnarvon died unexpectedly. A cut mosquito bite became infected, and Carnarvon, who suffered poor health, perished from pneumonia. His death marked the beginning of a decline in Carter's outlook of the excavation.

The Egyptian Antiquities Service, now unencumbered by a British Protectorate, began to exert greater control over the excavation site. Carter bristled against the diminishing power over his find.

The Mummy's Curse

In January 1923, Carnarvon had to minimize the press' intrusion on excavation workers, and he had to acquire additional funding for the expensive project. Carnarvon solved both problems by signing an exclusive contract with the *London Times*. Reporters from other publications resented the *Times'* monopoly on breaking news.

Lacking facts, excluded journalists eagerly reported that, at the time of Carnarvon's untimely death, Cairo's lights blacked out, while in England, Carnarvon's dog, Susie, howled and dropped dead. The mummy's curse slew Carnarvon, newspapers informed their readers.

Journalists backed up the curse story with reports of ominous hieroglyphs. One reporter invented a curse written in hieroglyphics on the door of the second shrine: "They who enter this sacred tomb shall swift be visited by wings of death."

In front of the Anubis shrine of Tutankhamun's tomb, Carter had found a wick lamp with a small mud base bearing hieroglyphics that read: "It is I who hinder the sand from choking the secret chamber. I am for the protection of the deceased." One correspondent embellished Carter's find by adding the words, "and I will kill all those who cross this threshold into the sacred precincts of the Royal King who lives forever."

Six of the 24 people present at the official tomb opening died by 1934. No matter how natural the circumstances, each death rekindled stories about the mummy's curse. Recent statistical analyses show that those present at the opening did not experience a decreased survival time.

In February 1924, Carter conducted special guests into the tomb for a long-awaited event: examination of Tutankhamun's mummy. It was not a simple matter. First, massive granite slabs of the sarcophagus lid had to be pried up, so that stones could be rammed into the opening. After securing straps around the lid, the granite slabs, weighing nearly two tons, were raised from the stone coffin. When Carter shined a light into the sarcophagus, he saw an object obscured by linen shrouds. He removed the linen wrappings to reveal a golden effigy of the boy king made of gilded wood and decorated with thin gold plates, faience and semiprecious stones.

After they left the tomb, Carter asked Pierre Lacau, director general of the Antiquities Service, if the excavators' wives could visit the tomb before the press viewing on the following morning. The next day, Carter learned that the Minister of Public Works had denied the request. Carter responded to the calculated insult impulsively. With the massive stones hanging over the young pharaoh's remains, Carter and the excavators went on strike.

The Egyptian authorities accused Carter of negligence. Carter demanded apologies from the government for the disrespect it showed him and his staff. Instead of an act of contrition, the government required Carter and Lady Carnarvon to sign a waiver stating that they would not make a claim on objects found in the tomb.

After a year of negotiation, the Egyptian government agreed to pay Lady Carnarvon £36,000, the approximate amount of Carnarvon's expenses over the years. Carter received about £8,500 of this sum and was allowed to resume work on the excavation.

In January 1925, Howard Carter returned to the Valley. He raised the lid of the gilded coffin that he had seen a year before and found a second coffin, this one covered with fine linen shrouds and adorned with garlands of flowers. Carter rolled back the shrouds to reveal yet another coffin, one fashioned of thick gold foil inlaid with engraved glass that simulated red jasper, lapis and turquoise.

After they pried open the lid of the third coffin, Carter saw Tutankhamun's mummy. A sticky, hardened, black resin covered the body and bound the king's head to a life-sized gold mask inlaid with blue glass simulating lapis lazuli.

For four days, the team unraveled bandages and recorded each of the artifacts hidden within the wrappings. With the assistance of Dr. Douglas Derry, professor of anatomy at the Egyptian University, Carter sliced through 13 layers of stiff linen. By the time that they had finished, they had collected 143 pieces of jewelry, ornaments, amulets and implements.

To examine the resin-coated mummy, they cut off the head at the neck and used hot knives to pry the skull from the mask. Then they separated the pelvis from the trunk and detached the arms and legs. Two medical specialists examined the mummy and concluded that the king had died between the ages of 18 and 22. After the examination, they reassembled the remains on a layer of sand in a wooden box with padding to conceal the damage and replaced the mummy in the tomb.

Howard Carter and colleague remove consecration oils that covered the inner coffin of King Tutankhamun, circa March 1926.

The Tutankhamun excavation marked Carter's last. He died in England in 1939.

In his account of the first exploration of the Antechamber, Carter wrote that, "The day following (November 26) was the day of days, the most wonderful that I have ever lived through, and certainly one whose like I can never hope to see again."

This might seem like an odd statement. According to his official account, Carter had the opportunity to explore the Burial Chamber and Treasury months later. What Carter failed to mention in his version of events is that he, Carnarvon and Evelyn had secretly revisited the tomb. After breaking through the Antechamber, they risked their license by exploring the Burial Chamber and Treasury. Afterwards, they had disguised traces of their adventure.

A report of their unauthorized exploration emerged years after Carter's death. Yet Carter might have hinted about the excursion in his book. "I think we slept but little, all of us, that night," he wrote.

CSI: Egypt

After Howard Carter and his team finished their examination, they replaced Tutankhamun's remains in his coffin. The mummy rested undisturbed for about 40 years.

In 1968, a group from the University of Liverpool used x-rays to learn more about the young pharaoh. They found bone fragments inside Tutankhamun's skull. Did political enemies bludgeon the young pharaoh to death? X-rays revealed that the mummy lacked the sternum and some frontal ribs. Had the pharaoh's chest been crushed in a chariot accident? The spine displayed signs of scoliosis.

Almost 40 years passed before researchers applied a new technology to the riddle of the boy king. In January 2005, scientists removed King Tutankhamun's mummy from the sarcophagus, and transported it to a nearby trailer equipped with a mobile computed tomography scanner. The machine scanned the body in 0.62-millimeter slices, producing 1,700 three-dimensional images. An Egyptian team of radiologists, pathologists, and anatomists, as well as three international experts examined the scans.

The images revealed a well-nourished 19-year-old boy who stood about five feet six inches tall and had a slight build. He appeared to have enjoyed good health, or at least, avoided any disease that would have left a trace on his remains. The experts decided that a misalignment during embalming had produced the spine's curvature, not scoliosis. Tutankhamun had a slightly cleft palate, an overbite and an elongated skull.

The scan revealed the bone fragments uncovered by earlier x-rays. However, the experts concluded that embalmers or Carter's group had inflicted the damage. The pharaoh had not been murdered by a blow to the head.

A break in Tutankhamun's left thighbone suggested a possible cause of death. The scan revealed a thin coating of embalming resin around a bone break that showed no sign of healing. This suggested that the pharaoh broke his leg just before he died. A fatal infection could have set in.

The damaged chest remains a mystery. Did embalmers remove the breastbone and part of the front rib cage?

Modern Embalming

Around 4400 BC, Egyptians embalmed corpses with techniques that varied depending upon the deceased's station in life. An embalmer might immerse the body in carbonate of soda, fill torso cavities with aromatic substances, and anoint the skin with oils. Embalming methods spread from Africa and Asia to Europe. During the Middle Ages, techniques for preserving European royalty included immersion in alcohol and insertion of preservative herbs into the body.

In the early 18th century, Dutch anatomist Fredrik Ruysch injected chemical solutions into the arteries of human remains to delay the decay of anatomical specimens. His formulations remain unknown.

Dr. William Hunter, a Scottish physician and anatomist of the 18th century, did not keep his arterial embalming techniques secret. He replaced blood with an injected solution of oil of turpentine, oil of lavender, Venice turpentine, oil of rosemary and vermillion. Hunter also removed organs from the body cavity, cleaned and injected them, cleaned the cavity, and replaced the organs.

Modern embalming emerged from the American Civil War. Thomas H. Holmes, who had a commission in the Army Medical Corps, injected the arteries of dead soldiers with a solution of zinc chloride and arsenic to prepare the bodies for burial. In a four year period, he claimed to have embalmed – for a fee – over 4,000 bodies.

In 1861, Holmes embalmed the corpse of military hero Colonel Elmer E. Ellsworth, a long-time acquaintance of President Abraham Lincoln. Display of the colonel's corpse at the White House and further venues inspired others to embalm bodies near battlefields, a practice that ensured the survival of Union soldiers' remains throughout the long trip home.

By the early 20th century, formaldehyde began to replace more hazardous embalming preservatives. North American embalmers continue to preserve bodies with formaldehyde solutions injected into arteries and the body cavity.

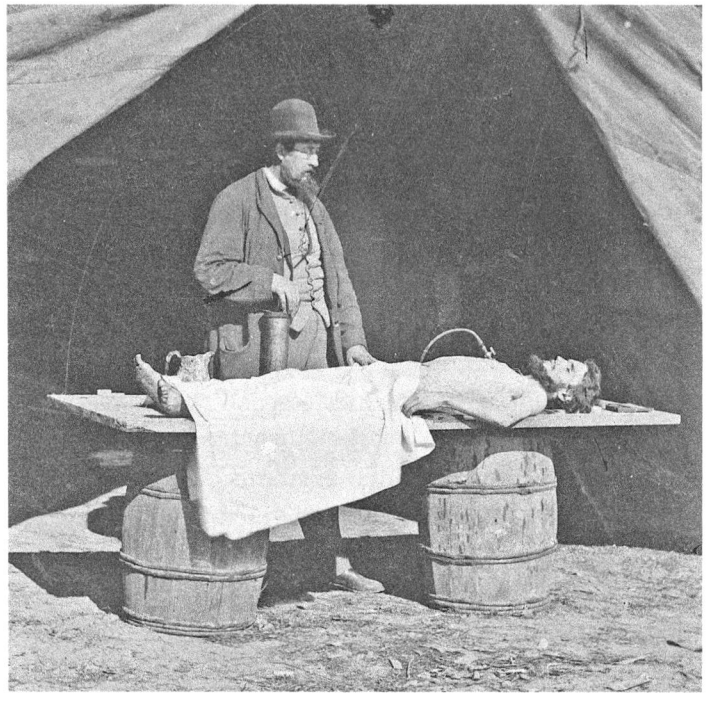

Embalming surgeon at work on a soldier's body, circa 1860-1865. Source: US Library of Congress.

The Timeless Appeal of Clocks

Time constraints and time pressures define modern life. Daily activities revolve around schedules, appointments, billable hours and deadlines in ways too time-consuming to mention. Despite complaints about lives run by the clock, humans have a compulsion to track time's passage, and have been doing so – for a very long time.

Early Timekeepers

About 5,000 to 6,000 years ago, people of the great civilizations in the Middle East and North Africa decided to track time within a day. The ancient Egyptians plunged a stake – or gnomon from the Greek "to know" – into the ground to carve a day into slices of time. Illuminated by the sun as it traversed the sky, the gnomon cast a shadow that changed in direction and length.

During the following millennia, the growing complexities of city life and trade necessitated a greater precision in tracking the passage of time. The gnomon became calibrated. A piece of a calibrated Egyptian sundial has been dated to about 1500 BC. The sundial had a T-shape with variably spaced marks etched into the long stem. One end of the stem was bent at a right angle so that the T's crossbar stood upright. In the morning, the long stem would be oriented east and west. With the upright crossbar on the east end, the arm cast a moving shadow over the stem's marks. At noon, the device would be turned in the opposite direction, so that the upright crossbar continued to cast a shadow on the calibrated stem.

By this time, the Egyptians had attached a special significance to the number 12, the number of constellations observed before the annual flooding of the Nile River. The T-shaped sundial divided daylight into 12 equal parts: 10 parts plus two twilight hours in the morning and evening. The duration of these temporal hours varied with the seasons' changing length of days. The Greeks adopted the use of temporal hours, as did the Romans, who spread the convention throughout Europe. Temporal hours remained a standard measure for more than 2,500 years.

In the search for improved accuracy, the Egyptians developed sundials that measured time solely by the direction of a shadow. The design of sundials transformed from traditional horizontal or vertical plates. The hemispherical sundial, for example, had a bowl-shaped depression inscribed with sets of time lines for different seasons.

As cultures became more complex, a need arose to track time on cloudy days, indoors, and at night. Calibrated candles, hour glasses and water clocks supplemented sundials. Each had drawbacks.

Historians credit the 9th century Saxon king Alfred the Great for introducing the use of burning candles to track time. In *A Child's History of England* (1852), Charles Dickens described how the king arranged his busy schedule.

> Every day he divided into certain portions, and in each portion devoted himself to a certain pursuit. That he might divide his time exactly, he had wax torches or candles made, which were all of the same size, were notched across at regular distances, and were always kept burning. Thus, as the candles burnt down, he divided the day into notches, almost as accurately as we now divide it into hours upon the clock.

During China's Sung dynasty (960-1279), time could be tracked by lighting graduated candles or sticks of incense. These blazing timekeepers had a limitation: Somebody had to light the next candle or piece of incense.

The hourglass tracked time by controlling the flow of sand from an upper chamber into one below. But abrasive sand eventually enlarges the connection between the chambers, and the flow rate increases. Producing a sandglass sufficiently large to last through the night posed another problem.

The water clock, or clepsydra for "water thief" in Greek, became the most successful supplement to the sundial. The earliest form of water clock might have been a bowl with a hole in the bottom that would be placed in a water-filled vessel. As water percolated through the hole, the passage of time could be read from calibrated marks etched inside the sinking bowl.

Another early form of water clock also consisted of a large bowl with lines etched into its inner surface and a hole placed near its bottom. In this version, the bowl would be filled with water that slowly leaked, causing the water level to fall from one time line to the next. This type of a clepsydra lacked reliability; the dwindling water supply could not provide a uniform water flow.

A more reliable clepsydra, the inflow clock, consisted of two basins. To maintain a stable downward pressure, water would be added continuously to a basin with a hole in its bottom. At a constant rate, water flowed from the leaky basin into a second basin marked with graduations for hours. The rising water level in the second basin marked off time's passage.

Greeks and Romans devised increasingly more elaborate water clocks. Some water clocks rang bells and gongs, some included figurines that moved through doors, and others displayed astrological models of the universe. Plato reportedly invented an alarm clock to rouse slumbering students in the academy that he founded in 378 BC. To signal the start of a new day, Plato's water clock dumped metal balls onto a copper plate.

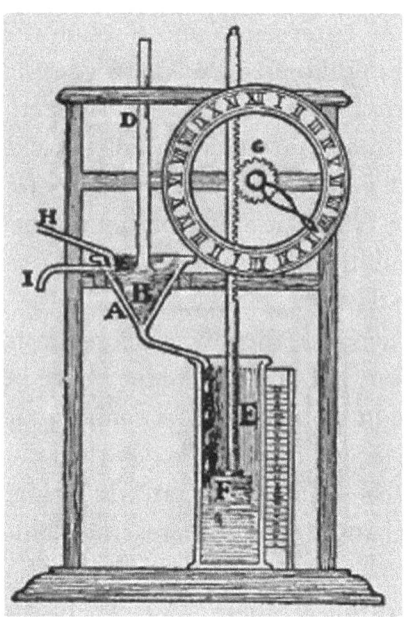

A Greek clepsydra. Water falls into the cylinder (E), causing the float (F) to rise and turn (G), which carries the hour hand. Source: Forman, S.E. *Stories of Useful Inventions* (The Century Company, 1914).

During 11th century China, Su Sung and co-workers built a 40-foot tall clock designed to reproduce the movements of the sun, moon and selected stars. A water wheel powered the massive apparatus, a wheel that turned at a steady rate determined by the filling of water buckets from a clepsydra.

Despite increasing sophistication, water clocks remained entrenched in the past. A water clock's 12 marks corresponded to the sundial's temporal hours. In northern latitudes, reliance upon sundial time required complicated adjustments. Slate fragments, unearthed in a Belgium abbey and dating to about 1267, explain how to set a water clock for each day of the year.

Until the late Medieval Period, Europeans relied upon sundials and water clocks for measuring time. Then, the mechanical clock made a timely appearance. It restructured daily life.

Mechanical Clocks: Revolutionary Concept

In medieval Europe, the Christian Church used temporal hours to designate times of prayer. Medieval water clocks primarily functioned to sound alarms for bell ringers, who would signal canonical hours.

Through the ringing of canonical hours, people developed an awareness of time. The activities of conducting business also created a sense of time, a sense that became abstracted from the sun's daily travel and from hours designated for worship. Merchants and monks alike wanted a more precise method for tracking the passage of time than provided by sundials and water clocks. The mechanical clock fulfilled this need.

The mechanical clock did not merely advance timekeeping performed by elaborate water clocks; it divorced time from a relationship with nature. Unlike the temporal hours of sundials and water clocks, the mechanical clock divided time into equal segments. The convention of a 24-hour day/night cycle has its roots in ancient Egypt, whereas the division of hours and minutes into 60 parts derives from the Mesopotamians. The sun's passage across the sky did not preordain these divisions. In fact, for several years after the French Revolution, the French experimented with decimal time. They had ten equal hours for each day/night cycle, one hundred minutes in each hour and one hundred seconds in each minute.

The introduction of mechanical timekeeping – "of the clock" or "o'clock" – transformed the concept of time from something that can be tracked to something that can be used. Mechanical time enabled the coordination of business, religious and military activities. It supported scientific progress by revealing a world of mathematically-defined events, and mechanical time spurred industrial change that began in 18th century England.

Old habits died hard. When mechanical clocks acquired dials, the clockwise motion of the hands mimicked that of a sundial's gnomon in the northern hemisphere. Early timekeepers tracked time with something that moved continuously: the sun's passage, the flow of water or sand, or the passage of fire burning a candle or a stick of incense. These systems reflect the idea that time appears to move continuously and should be tracked by something that flows as well. The first mechanical clocks also used a flowing movement: the fall of a weight.

Weight-driven Clocks

A mechanical clock has three critical elements: a source of power, a means to regulate the power, and a way to transform regulated power into an indicator of time. A falling weight must have appeared to offer a natural driving force for a clock, but it also presented a problem. A falling weight does not move at a constant speed; it accelerates as it falls. While a brake can slow the fall of a weight, a brake becomes worn and the speed of the weight's fall increases. The invention of the escapement mechanism solved the dilemma. It is this regulator of the clocks' driving apparatus that represents the greatest innovation of the weight-driven clock.

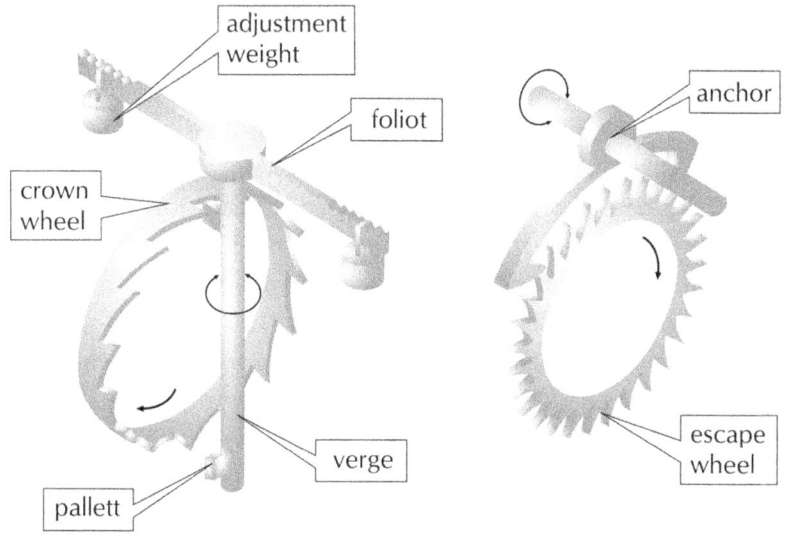

A verge escapement with foliot (left) and an anchor escapement (right). Courtesy of bricksandbrass.co.uk.

The escapement controls the escape of force from the falling weight by releasing equal amounts of energy at identical, short intervals. The verge-and-foliot escapement was the most widely used device for controlling the release of energy. The foliot, a short horizontal rod, carries balance weights at either end that can be positioned along the rod. Connected with the foliot, a vertical rod called the verge has two small plates, or pallets, protruding at about right angles to each other.

In a clock with this escapement, a toothed crown wheel slowly turns, driven by a falling weight fixed to the end of a cord wound around the wheel's axel. When the wheel turns, it engages the upper pallet of the verge. This stops the wheel's motion and makes the foliot oscillate. The oscillating foliot moves the verge, allowing the wedged crown wheel tooth to gradually force the upper pallet to move until the wheel escapes. Now free, the wheel advances until the lower pallet ensnares a tooth, causing the foliot to swing in the opposite direction. Aided by the oscillating foliot, the wheel gradually pushes the lower pallet until the wheel frees itself and advances once again. The crown wheel, which escapes from one pallet only to be caught by the other, turns one tooth at a time, producing the familiar tick tock sound of a mechanical clock.

The crown wheel turns at a rate that depends upon the falling weight and by the positions of the foliot's small weights, which determine the tempo at which the foliot swings back and forth. By arranging the driving weight and balance weights, a clock can be made to move with a preset, regular motion.

Weight-driven clocks appeared in cathedrals, churches and castles during the late 13th and early 14th centuries. These early mechanical timekeepers lacked a dial with hands; they signaled time with bells. The name "clock" for the new timekeeper originated from clocca, the Medieval Latin word for "bell."

According to legend, in 1370 King Charles V of France could no longer stand the cacophony of Paris' bells – some signaling canonical hours, some marking the beginning and end of a work day for certain professions, and yet others approximating equinoctial hours. Charles declared that all bells in Paris must ring simultaneously with the clock on the Royal Palace, which struck 24 equal hours. Although the story may be apocryphal, it symbolizes the replacement of the temporal hour with the fixed uniform hour.

By the middle of the 14th century, clockmakers had adapted the weight-driven mechanism to smaller clocks. People no longer relied upon the bells of a public clock; they could tell time at home.

The weight-driven mechanism did not offer an ideal method for tracking time: It limited miniaturization and required a clock to be vertical and stationary. Early weight-driven clocks also lacked reliability due to friction between hand-cut gears. However, it was the nature of the regulator that created the most significant barrier to dependability.

The regulator of a weight-driven clock performed two functions. The oscillating foliot defined segments of time, while the attached verge counted the segments by blocking and releasing the crown wheel. The foliot lacks a natural cycle of oscillation. Consequently, fluctuations in force transmitted by the crown wheel to the verge affected the foliot's rhythm. A reliable clock required an oscillator with its own inherent stable frequency, undisturbed by the other parts of the clock.

Mainspring Clocks

Early in the 15th century, clockmakers devised a substitute for gravity as a source of power: the unwinding of a tightly coiled spring, or mainspring. The conversion to spring power required a new trick. In a weight-driven clock, weights imparted an equal force to clock trains throughout the course of their fall. A spring's force, however, diminishes as it unwinds. Clockmakers had to find a way to keep the force constant.

The solution took the form of a cone-shaped, grooved pulley called a fusee wheel, positioned between the mainspring and the clock's wheel train. A cord or chain connects the fusee to a barrel that houses the mainspring. With a mainspring fully wound, the cord pulls on the cord-wrapped fusee at its narrow end, where the cord has relatively little leverage. As the mainspring winds down, the cord unwinds from increasingly wider parts of the fusee so that the increasing leverage compensates for the spring's diminishing strength. The combination of fusee and mainspring delivers constant power to the clock's gear wheels.

The fusee and mainspring enabled the development of the portable clock as well as the manufacture of the pocket watch. Watches became so small that they could be embedded in a dagger's hilt or in a finger ring. Worn primarily as status symbols, early watches lacked reliability. During the Renaissance, sundials had been devised that showed equal hours, and these sundials provided a standard of accuracy for mechanical clocks. Many watches came equipped with a sundial and a compass that the wearer could use to align the gnomon.

Poison IV?

With the notable exception of Big Ben, the vast majority of clock faces with Roman numerals indicate "four" with IIII, rather than the more commonly used IV. The reason for this is uncertain. Yet there are many theories.

The ancient Romans bear responsibility, according to one view. The Roman alphabet lacked J or U. Consequently, the letters I and V served as the first two letters in the name of the Roman god Jupiter, IVPITER. The Romans did not want to offend their god by using his abbreviated name, IV, on ordinary time-indicating devices. Sundials bore the politically correct IIII.

Some argue that the Romans used IIII simply because their numeric system relied upon addition, not subtraction. (The symbol IV indicates 5 (V) minus 1 (I).) It was during the Middle Ages that subtraction in the Roman numeral system came into vogue.

The choice, others suggest, relied upon economies of production. If four I's stand for "four," then 20 I's, 4 V's and 4 X's would have to be cast for a clock face. Conveniently, metal numbers could be cast using a mold that had a center rod with impressions for 10 I's, 2 V's and 2 X's balancing either side.

The simplest answer lies in aesthetics. The IIII on the right side of the clock face offsets the VIII on the left side. Use of IV would fail to create this symmetry.

In the long run, the use of IIII or IV doesn't seem to matter. In a psychological study, the investigator showed subjects a standard clock with Roman numerals. When asked to draw the clock from memory, the majority incorrectly showed IV, rather than IIII. They made this mistake even when warned in advance that they would have to draw the clock.

Pendulum Clocks

One day in 1583, the story goes, a teenaged Galileo Galilei became distracted by a swinging altar lamp while he attended prayers at the Cathedral of Pisa. Using his pulse to time the swings, he found that the lamp's swing took the same amount of time whether it made a wide arc or a small one.

This is so, because the time required for a pendulum to travel along an arc depends upon the pendulum's length and not on the distance traveled.

Although he never created a working model, Galileo had been aware that this principle of isochronism suggested the usefulness of a pendulum in a clock. In late 1656, Dutch scientist Christiann Huygens collaborated with clockmaker Salomon Coster to build the first pendulum clock. The clock used the same verge escapement that had been paired with the foliot for centuries. However, in Huygens' clock, the gravity-influenced motion of a pendulum replaced the mechanical oscillation of the foliot as the mechanism for determining beats of time. The pendulum's independent beat regulated the action of the verge escapement, resulting in a more reliable measurement of time.

In 1671, William Clement built a pendulum clock for King's College of Cambridge University. Clement's clock had a pendulum attached to an anchor escapement, a device that eliminated the verge with its pallets. As the pendulum swung, it rocked the anchor so that it caught and released each tooth of the crown wheel, which turned in a ratchet-like movement and drove the clock's hands.

The anchor escapement transmitted the natural regularity of the pendulum's beat to the face of the clock with greater accuracy and reliability. In a pendulum clock with a verge, the pendulum had to swing along a 20 degree path to turn the verge. While traveling such a distance, the pendulum followed a circular path, rather than the arc required for isochronism. This deviation introduced timekeeping errors. With the anchor escapement, the pendulum can make an arc of four degrees or less, a distance that decreases a variation from isochronism to a negligible amount.

A decade before Huygens and Coster assembled their pendulum clock, French monk and mathematician Marin Mersenne had calculated that a weight hanging on a 39.1 inch long piece of string would complete its swing in one second. Following the introduction of the Huygens-Coster clock, the Royal Pendulums incorporated the 39.1 inch pendulum and boasted a clock face with hands for hours, minutes and seconds.

The improved pendulum clocks with anchor escapements and seconds-beating pendulums varied by no more than ten seconds a day. The enhanced reliability of mechanical clocks overshadowed the performance of the sundial with its five minute range for a single reading. Eventually, conversion tables enabled an adjustment of a sundial's readings to clock time. Instead of serving as the standard of the clock, the sundial became standardized by the clock.

While the pendulum clock set a new touchstone for dependability, it did have a significant limitation as a timekeeper: The mechanism could not be adapted for timekeeping at sea or for pocket watches. According to one view, it was Christiaan Huygens who supplied a solution for accurate, portable timekeepers during the late 17th century. Instead of a pendulum, a spiral balance spring, or hairspring, regulated the clock's movement. The elastic spring transferred its periodic rhythm of compression and release to a balance wheel, a disk that rotates one way and then the other in a repeating cycle.

The combination of balance wheel and spring, still found in modern mechanical wristwatches, allowed 17th century portable watches to keep time to within ten minutes a day. Urban workers increased the demand for watches as the hourly wages of the factory system replaced cottage industries with their compensation on a piecework basis.

Improvements in pendulum clocks continued. In 1721, George Graham compensated for changes in the pendulum's length caused by temperature variations. He improved the clock's accuracy to one second per day. In 1889, Siegmund Riefler designed a clock that operated in a partial vacuum to maintain atmospheric pressure, temperature and humidity. It achieved an accuracy of a tenth of a second a day. By 1900, the Riefler Company of Germany crafted timekeepers accurate to within a hundredth of a second per day.

Twenty years later, English railroad engineer William H. Shortt demonstrated his so-called free-pendulum clock that kept time to within one second per year. His device included two pendulums connected by electromagnetic impulses. The record-breaking reliability of Shortt's clock soon would be eclipsed by the introduction of crystal time.

Electric Clocks

The mid- to late-19th century marks a period of vigorous efforts to harness electricity in the service of timekeeping. Five types of clocks exemplify how a timekeeper incorporates electricity and electromagnetism in its mechanism.

Electric clocks with pendulums driven by electromagnetic impulses. German clockmaker Mätthaus Hipp invented the most enduring method for energizing a pendulum with electromagnetic impulses. Clocks included his mechanism for more than 100 years. Hipp's 1842 clock had an electromagnet positioned below an iron-tipped pendulum. When the pendulum passed by the electromagnet, a magnetic pulse replaced lost energy. The pendulum of a typical mechanical clock provides the regulating oscillator. The Hipp clock pendulum serves as both oscillator and motor.

Electric clocks with pendulums driven by mechanical impulses. In these clocks, an electromagnet powers a mechanical force – basically, a swift kick – to maintain a pendulum's movement. For example, F. Hope-Jones' Synchronome had a toothed wheel that released a gravity arm that fell onto a pallet attached to the pendulum rod. After hitting the rod, the gravity arm touched an electrical contact that drove it back to the upper position, ready to fall again.

Clocks with electric winding. Clockmakers devised timekeepers that included an electric motor or an electromagnet to wind pendulums, balances, weights, or mainsprings. These types of electric clocks presented a challenge: ensuring continuous power during winding.

Secondary clocks. In this system, a master clock emits impulses to a secondary clock, which displays hours and minutes on a clock face. A secondary clock is a slave dial, not a true clock.

Synchronous clocks. The synchronous motor, patented by Henry Warren in the early 20th century, turns at a speed determined by power station generators, regulated to deliver a stable supply of alternating current. In a synchronous clock, the motor's movement transfers to a wheel train that controls the clock face. Like a secondary clock, a synchronous clock is really a slave dial. Beginning in the 1930s, clock manufacturers produced a type of synchronous timekeeper called a mystery clock. These devices had hidden mechanisms that moved hour and minute hands.

Quartz Clocks and Atomic Clocks

While working in his Paris laboratory in 1880, Pierre Curie and his brother Jacques observed the phenomenon that became known as the piezoelectric effect: An application of pressure to certain crystals generates an electrical voltage. Later, they confirmed that an alternating electrical current causes particular crystals to expand and contract at a steady rate.

In 1927, Canadian-born Warren A. Marrison, a telecommunications engineer at Bell Telephone Laboratories in New York, developed the first clock that used the regular vibrations of a quartz crystal in an electrical circuit. The accuracy of quartz clocks, achieving one or two thousandths of a second per day, far exceeded anything a pendulum clock can ever realize. A quartz crystal divides time by vibrating thousands, or multiple millions, of times a second. Mechanical clocks cannot gauge such minuscule slivers of time.

The early quartz clocks filled cabinets three meters high. Improvements in quartz crystal controllers enabled the Japanese Company, Seiko, to introduce the first electronic quartz wristwatch in the late 1960s. Within a decade, further technological advances transformed the digital quartz watch from a novelty to a serious contender for mechanical watches.

The modern digital watch contains a small battery, a capsule with a quartz crystal and electronic circuitry. Common watch crystals vibrate 32,768 times per second. An integrated circuit counts the quartz crystal's vibrations, and converts thousands of beats per second into data that can be transmitted to the clock face. Quartz clocks also inhabit personal digital assistants, cell phones and other electronic devices that reduce the demand for the venerable wristwatch.

While the quartz crystal clock serves for personal timekeeping, it cannot support the coordination of modern communications systems, financial transactions, navigation systems and power distribution. Quartz clocks rely upon a mechanical vibration with a frequency that depends upon a crystal's size and shape. Modern life depends upon frequencies generated by atoms.

In 1949, Harold Lyons and his associates at the National Bureau of Standards in Washington, D.C. built a molecular clock that measured frequencies emitted by ammonia gas. This device only slightly improved conventional clocks. Attention shifted to the cesium atom and its electrons, which oscillate between two energy states. In 1955, Louis Essen and Jack Parry built the first cesium atomic clock at the

National Physical Laboratory in England. About ten billion times more accurate than a pendulum clock, the cesium clock relies upon atoms that, unlike quartz crystals, do not fatigue and oscillate ceaselessly without distortion.

Studies with atomic clocks confirmed an observation made with quartz clocks: The Earth's rotation has unpredictable irregularities. In 1967, the 13th General Conference of Weights and Measures redefined the second as the duration of 9,192,631,770 periods of the radiation corresponding to the transition between two energy states of the cesium-133 atom. This transformed our standard of time from one grounded in the movement of the planet to one riding on the motions of the infinitesimally small.

Plastics

"I just want to say one word to you," Mr. Maguire says to Benjamin Braddock in the film, *The Graduate* (1967). "Just one word."

"Yes, sir," Benjamin says.

"Are you listening?"

"Yes, sir, I am."

"Plastics," Maguire tells Benjamin.

At that time, plastic symbolized artificiality, a facet of culture that Benjamin desperately wanted to avoid. Over the years, the negative connotation faded. Plastic endures. After all, plastic is much more than a useful material or even a cultural phenomenon. The invention and development of plastics reflect nothing less than the human urge to shape and mold the environment.

The Natural Plastics

The word "plastic" is derived from the Greek *plastikos*, or moldable. A plastic material is one that can be molded or shaped into different forms. In scientific terms, a plastic is a polymer, a large molecule composed of smaller units joined to form a long chain. The use of plastic materials predated knowledge about polymers by thousands of years. Although it may seem contradictory, the first plastics were natural materials.

Animal horn, composed of fibrous polymers of protein, provided one ancient form of natural plastic. Around 2,000 BC, artisans softened tortoiseshells in hot oil and reshaped the material into ornaments and food utensils. During the middle ages in England and Europe, horners softened pieces of cow horn in boiling water or in an alkaline solution. If a horner needed a thin sheet of material, he peeled off layers of horn along growth lines. A thick piece of horn would be fashioned by pressing thin sheets together. After achieving the desired thickness, softened horn would be squeezed into a mold. Horners made rustproof spoons, flexible combs, and intarsia to decorate wood. Taking advantage of the translucent and shatter-resistant qualities of horn, craftsman also fashioned windows for lanterns. By the 19th century, an industry was poised to sell mass-produced items molded from horn.

Shellac offered an alternative substance for making useful objects. Around 1290, Marco Polo returned to Europe with shellac. He discovered shellac in India, a part of the world where craftsman had used this natural polymer for centuries. A new European shellac industry rested upon the tiny shoulders of *Laccifer lacca*, a small insect native to India and Southeast Asia. Shellac production begins when a female lac insect inserts its proboscis into a twig or small branch of a soapberry, acacia, or a fig tree, and then consumes the sap. As the insect feasts, it secretes a thick liquid that dries, becomes hard and immobilizes the insect. After a male lac fertilizes the female, the immobilized insect continues to secrete the liquid until it is covered. The insect lays hundreds of eggs and dies. The young insects hatch, chew out of the covering, and strike out on their own.

The shell of the lac, or shellac, has many useful properties. After it has been cleaned and dissolved in alcohol, shellac can be used as a protective, almost transparent, coating for floors and furniture. When subjected to heat and pressure, shellac can flow into a mold. The addition of fibers to shellac increases its strength. By the end of the 19th century, a new industry produced shellac fashioned into knobs, buttons, electrical insulators, and phonograph records. The shellac industry diminished over time, but held ground until synthetic plastics became dominant in the 1930s.

In 1843, William Montgomerie returned from Malaysia with news about another type of natural plastic, gutta percha. Derived from the latex of *Palaquium gutta* trees, gutta percha can be softened in hot water and pressed into a shape. At room temperature, gutta percha is solid and does not easily break. The Gutta Percha Company, established in 1845, manufactured figures of animals, chess pieces, tea trays and inkstands.

Inventors discovered that gutta percha was not only an exceptional insulator, but could be extruded into long strips suitable for encasing a wire. The insulated cable was waterproof, flexible, and resistant to chemical degradation. Gutta percha insulation protected the first underwater telegraph cable, which ran under the English Channel from Dover to Calais. In 1849, the Morse Telegraph Company of the United States laid a gutta percha-insulated cable across the Hudson River. Gutta percha also protected the transatlantic telegraph cable laid in the 1860s. Synthetic plastics of the 1920s and 1930s diminished gutta percha's significance as an insulator.

Machines cover wire of the Atlantic telegraph cable with gutta percha at the Gutta Percha Company. Source: *The Illustrated London News* (March 14, 1857).

A limited supply numbered among the drawbacks of natural plastics. The problem of meeting demand became acute in the shellac industry. One pound of shellac could require the efforts of tens of thousands of tiny lac insects. The quantity and quality of a shellac harvest varied with rainfall, temperature, and the actions of predator insects. Natural plastics also suffered from deficiencies in their properties. Horn tended to curl over time. When shellac absorbed water, it blanched and turned brittle. Under dry conditions, shellac eventually darkened. Contaminants in a batch of gutta percha created areas that lost its insulating property; electric circuits short-ed out. Manufacturers sought an alternative for natural plastics, something that was not limited in supply and could be produced with a consistent quality. They needed more control over the production of plastic.

Modifications of Natural Plastics

Many tried to cure the deficiencies of horn, gutta percha and shellac by altering natural substances. Some attempted to create artificial horn from casein, a protein in milk curd. Dried milk curds would be ground into powder and mixed with water to make dough that could be molded into a desired shape. These items tended to dissolve when wet. In 1897, German printer Adolf Spitteler made a moldable plastic by treating casein dough with formaldehyde. Galalith, or milkstone, was used to make consumer products, such as buckles, buttons, fountain pen barrels, umbrella handles, and knitting needles. Casein-based plastic is still used to make buttons.

A gum rubber mixture seemed to offer another alternative to natural horn. In 1839, Charles Goodyear kneaded powdered sulfur into gum rub-ber and heated the blend. While small amounts of sulfur improved the properties of the gum rubber, the addition of large quantities of sulfur, amounting to 50 percent of a mixture, produced a rubber composition hard enough to shatter. The hard material, called Ebonite or Vulcanite, did not bounce like vulcanized rubber, but it could be used as substitute for horn.

In 1846, Swiss-German chemist, Christian F. Schönbein discovered that treatment with nitric acid and sulfuric acid converted the cellulose of cotton into an explosive. The volatile property required highly nitrated cellulose. Moderately nitrated cellulose, or pyroxylin, possessed very different prop-erties. Pyroxylin could be dissolved in an organic solvent. Evaporation of the solvent left a thin film of collodion, a hard, water-resistant material that seemed similar to horn.

British scientist Alexander Parkes mixed pyroxylin with oils to pro-duce a new type of plastic material that could be heated and molded. He introduced Parkesine to the world at the 1862 Great International Exhibition in London. The inventor claimed that Parkesine offered the qualities of India rubber, ivory, tortoise-shell, horn, and gutta percha, and would replace those materials. He founded a company in 1866 to exploit Parkesine. However, the high cost of the raw materials doomed the company.

In 1863, American inventor John Wesley Hyatt experimented with pyroxylin. He planned to win a $10,000 reward for an ivory substitute useful in the manufacture of billiard balls. At first, Hyatt made a few billiard balls out of shellac and wood pulp, and found that the mixture lacked ivory's elasticity. Hyatt tried a different approach. He mixed powdered pyroxylin with pulverized gum camphor, wet the blend to disperse the powders, and dried the composition. He then placed the mixture in a mold, heated it and pressed it into a shape. Hyatt named his new material celluloid.

Celluloid billiard balls had a flaw: They shattered upon impact. Celluloid dental plates also proved a failure. John Hyatt and his brother Isiah achieved success with celluloid combs and mirror frames produced by their Celluloid Manufacturing Company. Between 1873 and 1880, the Hyatt brothers initiated licensing arrangements with companies that made celluloid harness trims, baby rattles, combs, piano keys, shirt collars and cuffs, surgical instruments and other small items. Celluloid is still used to manufacture ping-pong balls.

The diverse applications of celluloid revealed the potential of an artificial plastic material. Inventors pursued a new type of plastic, one that they made from scratch.

Advertisement for Ivory Pyralin, DuPont's trade name for celluloid. Source: *DuPont Magazine* 12(2):17 (1920).

Bakelite, the First Synthetic Plastic

Nineteenth century chemists experienced failed chemical reactions that produced a gunk, an insoluble, gummy resin. In the early 20th century, chemist Leo Hendrik Baekeland created a gunk. Yet he did not see the result as a failure. Baekeland turned the blob into a new material that ushered in the Age of Plastics. *Time* magazine would salute the new plastic as "a composition, born of fire and mystery."

In 1889, Baekeland and his wife Céline sailed from their native Belgium on their honeymoon. After visiting New York, they decided to stay. Baekeland invented a photographic paper that could be printed under artificial light. Velox paper captured the attention of George Eastman, who reputedly paid one million dollars for rights in the invention.

Financially independent at the age of 36, Baekeland moved his family to Snug Rock, a New York estate that overlooked the Hudson River. He converted stables into a private laboratory and turned his attention to developing synthetic shellac. The emerging electrical industry created a demand for insulating material that outstripped supplies of natural shellac.

Baekeland and an assistant started their search for a shellac substitute by experimenting with phenol and formaldehyde. For half a century, chemists had reported the production of a gummy formaldehyde resin with no apparent value. After three years of research, Baekeland found a way to control the pressure and temperature of the phenol-formaldehyde reaction with his Bakelizer, an egg-shaped vat that was part pressure cooker and part boiler. Over the next several years, Baekeland refined the new plastic compound, polyoxybenzylmethylenglycolanhydride.

The chemist introduced his plastic, now dubbed Bakelite, at the 1909 meeting of the New York chapter of the American Chemical Society. The audience greeted the invention with a standing ovation in recognition of the revolutionary nature of Bakelite, a moldable material that was neither a natural plastic nor a chemically modified natural material. Bakelite was totally synthetic, produced under conditions that ensured a purer and more uniform compound.

Bakelite appeared at just the right time. Diminishing supplies of shellac, ivory and rubber could not meet manufacturers' demands. Bakelite not only offered an inexpensive substitute for traditional materials; it had properties that surpassed these traditional materials. Bakelite resisted chemical degradation, heat, sea water and dry rot. It was shatterproof, electrically resistant, and did not crack or discolor upon exposure to sunlight.

Baekeland founded the General Bakelite Company to promote "The Material of a Thousand Uses" – an accurate description, rather than an idle boast. The electrical industry enthusiastically adopted Bakelite to produce fuses, sockets, switches and other components. Bakelite replaced shellac, rubber, and gutta percha as an insulator. The budding automobile industry incorporated Bakelite into door handles, distributor caps, radiator caps, gearshift knobs, instrument panels and steering wheels.

Bakelite consumer goods populated homes in jewelry, radio casings, toothbrushes, poker chips, teething rings, pipe stems, billiard balls, lamps, phonograph records, vacuum cleaners, buttons, cameras, telephones, kitchen appliances, and pen barrels. The resin, which could be produced in a variety of colors and molded into any shape, resonated with designers during the Art Deco movement of the 1920s and 1930s. Bakelite firmly established plastic products as a part of everyday life.

In time, Bakelite became a victim of the industry that it had spawned. A new generation of synthetic polymers would be produced with less cost and boast innovative, useful properties.

The Age of Plastics Matures

The 1930s saw the "poly" era of plastics, as polystyrene, polyethylene, polyvinyl chloride, polymethyl methacrylate, and other polymers underwent commercial development. New synthetic polymers invented between the two world wars marked a departure in the development of plastics. Traditionally, inventor-entrepreneurs sought a substitute for a particular material high in demand and low in supply. Now, researchers, typically affiliated with a company, started with a chemical discovery and devised a way to exploit it in the marketplace.

Two of the signature plastics of the time were polyethylene and nylon. E.W. Fawcett and R.O. Gibson of the Imperial Chemical Industries Research Laboratory (United Kingdom) discovered polyethylene by accident in 1933. They were experimenting with chemicals under high pressure when their pressurized container sprang a leak. Inside, they found a white, waxy material. Instead of counting the test as a failure, Fawcett and Gibson repeated the experiment with a deliberate release of pressure. The waxy substance was a new type of plastic. Three years later, Imperial Chemical Industries launched a large-scale production of polyethylene.

The Second World War brought military applications for polyethylene, including the use of the plastic as an insulator that enabled the installation of radar in airplanes. These lightweight radar systems allowed Allied aircraft to detect German bombers flying at night or through concealing thunderstorms. After the war, polyethylene found many uses in consumer products, including drink bottles, food storage containers and grocery bags.

During the late 1920s, Wallace Carothers investigated polymer chemistry at DuPont's Experimental Station in Wilmington, Delaware. By 1930, he had confirmed a theory that polymers are large molecules with a repeating structure. Carothers' team wanted to produce a "superpolymer" and decided that the key to constructing longer molecules was to remove water during the chemical reaction. Carothers and Julian Hill invented a molecular still, a device that enabled the manufacture of long chain polymers. In 1935, they invented a polymer that could be stretched into thin, strong fibers. They considered Fiber 66 to be a substitute for silk in silk stockings. The name of the new material endured a series of changes. "Norun" was abandoned, because the company did not want to warrant that the stockings would never run. The name was reversed to "nuron," a name rejected, because it sounded like neuron and suggested a nerve tonic. Nulon became nilon and finally nylon.

The first nylon stockings appeared for sale in October 1939, and then disappeared from the market; the outbreak of the war brought demands for military applications. The armed forces used nylon for sutures, mosquito netting, boat towlines, shoelaces for army boots, tents, parachutes, and tire cords in heavy trucks and aircraft.

The war transformed the plastics industry. In 1939, the United States consumed 592,000 tons of crude natural rubber; 98% came from Asia. By the end of 1941, the Japanese government had blocked rubber supplies from the Far East at a time when the United States' entry into the war created an extraordinary demand for rubber. Crash programs to create synthetic rubber led to extensive research into polymers that enabled the production of new plastics suitable as substitutes for metal and rubber. When military men returned to civilian life, they brought an appreciation for synthetic plastic. The 1950s experienced the use of polypropylene and other new plastics as well as improved pre-war plastics. Novel and enhanced plastics competed with traditional materials.

Today, North America produces about 24 per cent of the polymers manufactured around the world. The United States alone produces over 100 billion pounds of plastics each year. As plastic products load landfills, create a huge "Trash Vortex" in the ocean, and concern grows about the effects of plastic on the environment, researchers have turned to the development of biodegradable polymers. One biodegradable plastic, polyhydroxybutyrate, has been produced from bacteria and may offer an environment-friendly alternative to petroleum-based plastic. The properties of polyhydroxybutyrate have been improved with the application of nanotechnology, a development that highlights the ever-changing nature of a truly plastic industry.

Darwin Aboard the *HMS Beagle*

Cambridge University student Charles Darwin looked forward to finishing his last term during the spring of 1831. Alexander von Humboldt's *Personal Narrative* volumes (1799-1804), a subjective account of a scientist-explorer's adventures in New World tropics, had fired Darwin's imagination. "All the while I am writing now my head is running about the Tropics," Darwin wrote to his sister Caroline in a letter dated April 28, 1831. "In the morning I go and gaze at Palm trees in the hot-house and come home and read Humboldt: my enthusiasm is so great that I cannot hardly sit still on my chair. . . . I have written myself into a Tropical glow."

Anticipating a trip with Cambridge friends to Tenerife, the largest of the Canary Islands, Darwin learned Spanish. He also began a systematic study of geology. During the summer, Darwin accompanied Reverend Adam Sedgwick on a geologic field trip through Wales. When Darwin returned to Shrewsbury on August 29th for two weeks of shooting with his uncle Jos, he found a letter from John Henslow. Darwin's former mentor had been asked to recommend a young man with an interest in science and natural history to join a survey ship, and he thought that Darwin perfectly fit the requirements.

Darwin accepted the offer for a two-year survey. It became a five-year journey that radically altered his life.

Early Evolution of Charles Robert Darwin

Charles Darwin was born in Shrewsbury, Shropshire, England, on February 12, 1809. His parents were Susannah Wedgwood and Robert

Waring Darwin, an eminent physician. Charles had two famous grandfathers: pottery industrialist Josiah Wedgwood and Erasmus Darwin, a leading physician, prolific inventor and a philosopher who devised a theory of evolution. After suffering frequent illnesses, Susannah died in 1817, leaving Charles in the care of his three older sisters and his father.

One year after his mother's death, Charles was sent to board at Shrewsbury School in the center of town. The rote learning of classics sparked little interest in Charles, an attitude reflected in his lack of progress. Robert removed his son from the school and sent him to Scotland's Edinburgh University, where Charles would be a companion for his older brother Erasmus until Charles could attend medical school.

The study of medicine failed to fire up Charles' enthusiasm. He hated human anatomy classes – the dismembered corpses, reek of preservative, and even his anatomy instructor, Alexander Monro. In his autobiography, Charles wrote that Monro "made his lectures on human anatomy as dull, as he was himself, and the subject disgusted me." The fact that human anatomy classes had become associated with body snatching did not raise Charles' opinion about the subject.

He escaped from medical studies to natural history, his boyhood passion. Charles learned to identify rock strata, plants and animals, and practiced basic taxidermy. Professor Robert Edmond Grant interested Charles in the evolution theories of Jean-Baptiste Lamarck and Erasmus Darwin. Grant had his own hypothesis about evolution: that plants and animals shared a common marine ancestor.

Charles' inattentiveness to medical science did not please Robert Darwin, who accused his son of having an interest in nothing but game shooting, dogs and catching rats. In 1828, Robert uprooted Charles from Scotland and transplanted him at Christ's College in Cambridge. Destined for a career in the church, Charles soon became a protégé of Reverend John Stevens Henslow, a botanist.

Around this time, 26-year-old Captain Robert FitzRoy began looking for a gentleman to accompany him on a trip aboard the rebuilt brig, *HMS Beagle*. FitzRoy would sail the *Beagle* across the Atlantic to Bahia (Salvador) in Brazil to settle a disagreement between British and French measurements of the line of longitude. Then he would survey the coastline. While in Tierra del Fuego, FitzRoy would set up an Anglican mission, which provided the original motivation for the expedition. He planned to deliver a volunteer missionary from London and to return three English-speaking Fuegians to their homeland. Next, the crew would take chronometric measurements at certain locations in the Pacific and Indian oceans.

As soon as Charles learned about the opportunity to sail on the *Beagle*, he told his father. At first, Robert refused to let his son waste time on yet another diversion from a career. With his uncle's support, Charles convinced his father that he should go. Charles rushed to London, met FitzRoy and accepted the offer.

By December 3, 1831, Charles lived aboard the *Beagle*, moored in Plymouth. The captain assigned Darwin quarters in the chartroom, a nine by eleven foot room located at the stern of the ship. The cramped room held a bookshelf, oven, cabinets, a wash stand, and a four by six foot chart table; the mizzenmast also passed through the room. Charles, who shared his quarters with two of the survey officers, slept in a hammock suspended over the chart table and two feet from the ceiling. Before climbing into his hammock, he had to remove a drawer from a wall to make room for his feet.

Under clear skies, the 90-foot ship with its 73 man crew sailed from Plymouth on December 27. Darwin immediately became seasick, a malady that plagued him the entire trip.

The Voyage of the *HMS Beagle*.

A Budding Clergyman Explores the New World

On January 6, the *Beagle* docked at Santa Cruz on Tenerife Island. Darwin had anticipated an exploration of the Canary Islands since reading Humboldt's book. At the time, however, England experienced an outbreak of cholera, and the crew had to stay on board for a 12-day quarantine. The captain did not want to delay his mission, so he set sail.

The *Beagle* arrived at Santiago in the Cape Verde Islands on January 16. Now, Darwin had the chance to leave the ship and explore. For the first time, Darwin saw tropical plants and animals. He became fascinated with an octopus that changed its color like a chameleon. A white band of shells buried in the face of a cliff interested Darwin. The shells had been raised about 45 feet above sea level, which supported Charles Lyell's theory that the Earth's features had been slowly changing over great periods of geological time. The idea contradicted a popular notion that the world was only 6,000 years old and unchanging, just as plant and animal species had remained invariable since the dawn of creation.

After a stay of 23 days, the *Beagle* headed for Salvador, Brazil, where Darwin explored rain forests on long walks by himself. In April, the *Beagle* moored at Rio de Janeiro. Darwin and a local English merchant trekked 150 miles to Rio Macao. Darwin returned with a collection of insects and plants, which he preserved. That summer, Darwin dispatched his first shipment of specimens and notes to Henslow.

FitzRoy surveyed the Patagonia coastline in August, giving Darwin a chance to collect fossils, such as the remains of extinct ground sloths. As far as Darwin could tell, many of the fossils represented unknown animal species. The captain complained that Darwin kept hauling useless junk on board his ship. By the end of the year, Darwin sent a second shipment of specimens and notes.

In March 1833, the *Beagle* anchored at the Falkland Islands, which Darwin described as "an undulating land, with a desolate and wretched aspect." Darwin found more fossils on the islands and he noticed two types of adaptations: one behavioral and one physical. On the islands, the upland goose did not fear humans, whereas the same species on the mainland had been hunted for generations and avoided humans. Darwin also observed a steamer duck with an unusually powerful beak that crushed shellfish. He decided to compare the fossils, plants and animals that he saw during the rest of the journey.

After the *Beagle* docked at the Rio Negro in August, Darwin explored the Pampas with gauchos. It was with these local cowboys that Darwin ate greater rhea, a large flightless bird that reminded him of an ostrich. Darwin heard about the smaller rhea, which lived south of the Rio Negro. The birds belonged to two different species, and Darwin wondered if they shared a common ancestor.

The *Beagle* sailed past the east coast of South America and around Cape Horn to the Pacific Ocean during the spring of 1834. Darwin continued to preserve plants and animals and to send specimens, fossils and notes to England. While he sailed around the world, Darwin's reputation as a naturalist grew back home.

On February 20, 1835, a powerful earthquake struck Valdivia, Chile, while Darwin rested in a nearby forest. The earthquake and resultant tidal wave destroyed the city of Concepción. FitzRoy tried to anchor his ship at Concepción and Darwin was left at the island of Quiriquina. Here, he saw that the earthquake had raised a bank of mussels about eight feet above the sea. He interpreted this as evidence that South America was slowly rising, which again supported Charles Lyell's theory that land masses rose over an extremely long time period. A month later, Darwin explored the Andes and discovered petrified trees in sandstone. The fossilized trees resembled those he had seen at sea level, 7,000 feet below, indicating to Darwin the possibility that a series of earthquakes had raised the mountains. Darwin accepted the idea that the Earth must be very old, unstable and changing.

In the fall of 1835, the *Beagle* arrived at the islands of Galápagos. "The constitution of the whole is volcanic," Darwin wrote in *Voyage of the Beagle* (1839), his journal of the trip. "With the exception of some ejected fragments of granite, which have been most curiously glazed and altered by heat, every part consists of lava, or of sandstone resulting from the attrition of such materials."

The natural history of the islands fascinated Darwin. "It seems to be a little world within itself," he wrote, "the greater number of its inhabitants, both vegetable and animal, being found nowhere else." The iguanas captivated Darwin, especially one type that swam and fed in the ocean. Well adapted to its environment, the marine iguana could stay underwater for at least an hour, using a flattened tail to swim. In contrast, the Galápagos land iguanas had limbs and strong claws adapted for crawling over the jagged terrain.

Darwin discovered plants and animals with unique physical variations that seemed related to species that he had seen on the mainland. Many of the animals had coloration that blended with the surrounding lava fields, such as dusky-colored species of birds that had brightly colored counterparts on the mainland. He also found tree-sized plants related to daisies and sunflowers. Eventually, he wondered if species of plants and animals from the mainland had reached the islands and then adapted to the new environment.

After his return to England, specimens of Galápagos birds with significant differences in beak shape and size would illustrate the mutability of a species. When he had collected the birds, Darwin thought that he had a mixture of wrens, finches, and gross-beaks. Later, ornithologist John Gould revealed that the birds were all ground finches with different adaptations.

January 1836 brought Darwin to Australia, where he found the platypus and unique marsupial animals. Darwin suggested that a different type of creation must have occurred in this part of the world. By early summer, the *Beagle* sailed around the Cape of Good Hope. Instead of continuing north to England, FitzRoy took the ship back to South America to take more longitude measurements, a detour that frustrated Darwin. "This zig-zag manner of proceeding is very grievous," he wrote in a letter to his sister Susan. "It has put the finishing stroke to my feelings. I loathe, I abhor the sea, & all ships which sail on it."

As the *Beagle* finally sailed toward England, Darwin finished his 770-page diary, and organized about 2,000 pages of notes and 12 catalogs of his specimens. Darwin also reminisced about his five-year journey. "And what are the boasted glories of the illimitable ocean?" he wrote in *Voyage of the Beagle*. "A tedious waste, a desert of water." Just as well then, that Darwin had only spent a combined 18 months at sea during his five-year journey. Still, Darwin recommended to naturalists the benefits of a long excursion – on land, if possible.

The *Beagle* anchored at Falmouth, England on October 2, 1836. Darwin returned to his family a changed man. After a five-year absence, Robert Darwin greeted his son by turning to Charles' sisters and exclaiming, "Why, the shape of his head is quite altered."

A Natural Historian Returns Home

Upon his return, Darwin made another discovery: His shipments of specimens and notes had turned him into a science celebrity. Darwin settled in London and presented a paper to the Geological Society on the formation of the Andes. He also married his first cousin Emma Wedgwood.

While living in London, Darwin started his first notebook on transmutation, or evolution. He drew a crude evolutionary tree with ancient life forms at the bottom and descendants branching off along the trunk. An evolutionary tree contradicted a popular view that species were unrelated and had remained unaltered since their creation. He wondered what caused species to change.

Darwin found one engine for evolutionary change in Thomas Malthus's "Essay on the Principle of Population," which argued that competition for food or land was a constant force that kept the human population in check. The same principle could apply to plants and animals, Darwin theorized. In the struggle for survival, an organism with a variation that offered a competitive edge would be more likely to leave offspring than those that lacked the advantage. Over time, more members of the species would have the useful trait and the species would alter through this natural selection.

In 1842, Darwin and his family moved to a village in the English countryside. By now, he had formulated the basis for his *Origin of Species*. He also appreciated the most dangerous aspect of his ideas, one that destroyed the view of humans as outside and above nature: the evolution of humans and apes from a common ancestor.

For almost two decades, Darwin refined his theories and sought evidence in changes of domesticated animal and plant species. He kept his ideas secret to avoid ridicule that had been aimed at other scientists who proposed their theories of evolution. Eventually, he did tell his secret to a few friends. In January 1844, he wrote to botanist Joseph Dalton Hooker, "At last gleams of light have come, & I am almost convinced (quite contrary to opinion I started with) that species are not (it is like confessing a murder) immutable." Darwin's theory of natural selection, which required the deaths of many to establish advantageous traits, did appear to murder the idea of a benign deity.

In June 1858, news about Alfred Russel Wallace's theory of evolution by natural selection stirred Darwin into action. Facing the loss of priority, Darwin wrote his book, *On the Origin of Species by Means of Natural Selection, or the Preservation of Favoured Races in the Struggle for Life*, which was published about one year later. After controversy settled, he published *The Descent of Man, and Selection in Relation to Sex* (1871). Debates about the evolution of humans continue to this day.

Darwin wrote a book for his family in 1876; he called it *Recollections of the Development of My Mind and Character*. His son, Francis, included the autobiography in *The Life and Letters of Charles Darwin*, published five years after Charles Darwin's death in 1887. "The voyage of the *Beagle* has been by far the most important event in my life, and has determined my whole career," Darwin wrote in his memoir. "I have always felt that I owe to the voyage the first real training or education of my mind."

Gyrocompass

In 1851, French physicist Jean Bernard Léon Foucault found a way to demonstrate the Earth's movement around its axis. He invented an improved version of the gyroscope, a wheel that maintained its axis in parallel to the Earth's axis while turning within a concentric set of rings. Before long, Foucault made an interesting observation: A spinning gyroscope initially aligned toward the north continued to point north. This characteristic was soon put to practical use.

During the late 19th century, the innovation of steel-hulled ships endangered the usefulness of the magnetic compass, which sailors had used for 700 years. The metal hulls not only shielded the compass from the Earth's magnetic field, but also created a false magnetic field. Entombed in a metal hull, the magnetic compass proved unreliable. Interference with the compass' function proved worse in new submarines. Inventors realized that a device like a gyroscope might replace the traditional magnetic compass.

Around 1900, German scientist Hermann Anschütz-Kaempfe planned to travel to the North Pole by submarine. Faced with the need to find an alternative to the magnetic compass, he experimented with gyroscopes. In 1908, he developed a geographically northward pointing gyrocompass that the German Navy soon adopted. Yet Anschütz-Kaempfe's gyrocompass had a drawback; it had to be mounted to eliminate the sea's motion.

In 1907, US inventor Elmer Ambrose Sperry designed his own gyrocompass that dampened oscillations from a ship's motion. After several tests, the US Navy accepted Sperry's gyrocompass and began installation with the battleship *U.S.S. Delaware* in 1911. By the time that the United States entered the First World War, many of its ships carried Sperry's gyrocompass.

The Shocking History of Electricity

The ancient Greeks knew a little about magnetism. By 800 BC, texts described magnetite, or lodestone, an iron oxide mined in the province of Magnesia in Thessaly. In 600 BC, Thales, an astronomer in the Ionian port of Miletus, discovered that, if he rubbed a piece of amber with a cloth, the fossilized resin attracted feathers and other light objects. The attractive power of energized amber seemed similar to the ability of lodestone to attract iron.

Philosophers and scientists eventually understood the connection between the forces associated with lodestone and amber. This insight would enable the generation and harnessing of electricity to fuel society. It took more than 2,000 years.

From a Frictional Charge to the Electrochemical Battery

Magnetism attracted much interest during the reign of Queen Elizabeth I. The Queen wanted to improve navigation, and particularly, compass reading. William Gilbert, who would serve as the Queen's chief physician, spent years studying magnetic phenomena. He tested amber, rock crystal, sulfur, glass and other materials for the ability to attract objects after energizing with friction. Gilbert distinguished between the attractive properties of magnets and the "electric force" created with amber and similar materials. He derived the word "electric" from *elektron*, the Greek word for amber. Gilbert's *De Magnete, Magneticisque Corpibus et de Magno Tellure* (*On the Magnet, Magnetic Bodies, and the Great Magnet of the Earth*), published in 1600, became the standard text on magnetism and electric force.

Otto von Guericke, the burgomaster of Magdeburg, combined his interest in Gilbert's studies with his fascination about the nature of space and vacuums. By 1670, he had invented a vacuum pump. He used the pump to evacuate air from two copper hemispheres, causing the surrounding air pressure to seal the two halves into a sphere. When he rotated the sphere and rubbed it with his hand, the sphere acquired attractive properties and produced sparks. He had created the force that Gilbert had called electric.

A later model of Von Guericke's static electricity machine had a solid sulfur ball the size of an infant's head. Von Guericke mounted the sulfur ball on a shaft that passed through the ball's center. One end of the shaft had a crank that he used to rotate the ball. Rubbing the sulfur sphere with a cloth or dry hands energized the ball with friction. An improved version included a belt-driven device to turn the sphere faster, and produce greater quantities of electricity. The machine made it possible to perform new studies into the nature of electricity.

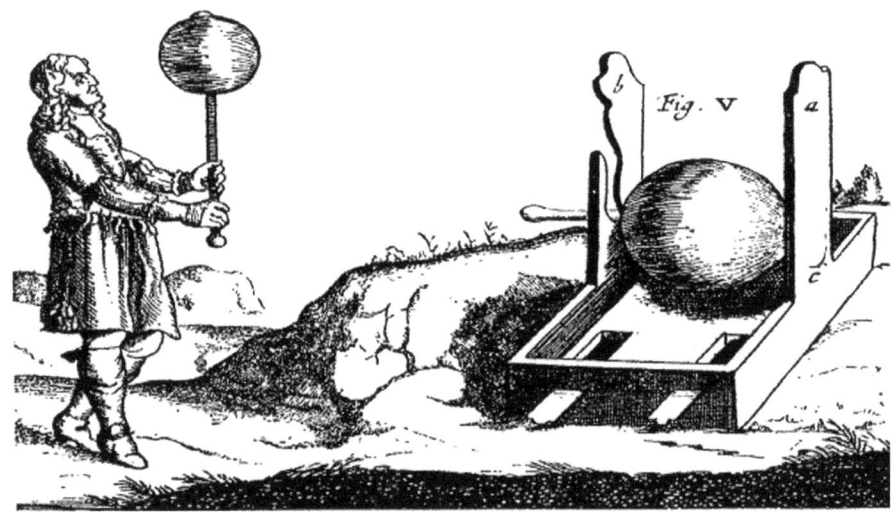

The illustration shows Von Guericke's electrical machine and sulfur globe (somewhat larger than an infant's head). Source: Durgin, William A. *Electricity: Its History and Development* (A.C. McClurg & Co. 1912).

Around 1728, Stephen Gray, a pensioner at Charter House in London, experimented with electricity, energizing silk, linen, wool, and other materials with friction. He discovered that, when he touched an electrified glass tube to one end of a metal rod, an object attached to the other end of the rod became electrified. Gray realized that electricity can be transmitted. Using a hemp thread insulated with silk, he sent an electric charge about 500 feet. Gray also showed that electricity can pass from one person to another, a discovery that led to the popular diversion of electrified kisses.

Gray made yet another discovery. Using silk lines, he suspended an iron rod with two pointed ends. When he approached the rod with an electrified glass tube in the dark, he produced cones of light. A method of generating light from electricity would be realized a century later.

By the mid-1700s, electricity had become a popular subject for study and amusement. Many wealthy families bought devices that allowed aristocrats to produce miniature lightning flashes from their fingers. Although static electricity generators became more efficient, the limited yields of electricity restrained experimentation. Two scientists independently solved the problem of storing electricity: Ewald G. von Kleist of Kammin, Pomerania in 1745 and Professor Pieter van Musschenbroek of Leyden, Holland in 1746. Although von Kleist could claim priority on the invention, the device became known as the Leyden jar.

One standard type of Leyden jar was made from a water-filled container of thin glass. A metal coating covered the lower two-thirds, both inside and out. A brass wire, capped with a small brass ball, was inserted through a cork lid and into water. An experimenter charged the jar by transmitting electricity from an electrostatic machine to the brass ball. Depending upon storage conditions, a Leyden jar could keep a charge for several days. A touch of the brass ball removed the stored electricity.

An energized Leyden jar stored a significant charge, as Musschenbroek discovered. "My right hand was struck with such force that my whole body quivered just like someone hit by lightning," Musschenbroek warned. "The arm and the entire body are affected so terribly I can't describe it. I thought I was done for." Despite the admonition, or perhaps inspired by it, others soon fabricated their own Leyden jars to experience the shock for themselves.

In the United States, Benjamin Franklin concocted a way to test the notion that lightning was composed of electricity. According to legend, a September 1752 thunderstorm offered an opportunity. Franklin attached a foot-long piece of metal wire to the top of a kite constructed with silk handkerchiefs and cross-sticks. He outfitted the bottom of the kite with a hempen string attached to a ribbon of silk, and affixed a key to the place where hemp and silk met. When Franklin and his son flew the kite in an approaching storm, they saw the hempen string's loose threads rise as if charged with electricity. The key produced a spark when Franklin touched it. From a safe distance, Franklin and his son charged a Leyden jar with electricity that flowed through the kite's wet string.

By storing an electrical charge, the Leyden jar played an important role in facilitating studies of electricity. But the device had a limitation: The jar produced only one discharge.

Alessandro Volta, a professor of physics at the University of Pavia, harbored doubts about a new type of force called animal electricity. Luigi Galvani had discovered animal electricity when his metal tools caused frog legs to twitch. The metal released a type of electrical flow from animal tissue, Galvani asserted. Volta repeated the experiment, and became convinced that moist conditions and the use of dissimilar metals in the probes generated electricity. The electricity was metallic, not animal in nature.

Volta tested various metals with an unusual instrument: his tongue. He placed combinations of silver, tin, brass, iron, and other metals in his mouth. A bitter sensation, he theorized, might be caused by a current that flowed from metal to metal via saliva. Based on these experiments, Volta assembled a device containing copper and zinc disks separated by pasteboard soaked in saltwater. The "voltaic pile" produced a continuous flow of electricity and sufficient electric current for study. Unlike the Leyden jar, the voltaic pile's charge was constantly renewed. Volta had invented the first battery.

At London's Royal Institution, chemist Humphry Davey wanted more power; he constructed progressively larger batteries. By 1808, he had fabricated a gigantic battery of 2,000 pairs of plates. Davey's experiments convinced him that electrochemical interactions produced electricity in the voltaic pile. In 1807, he showed that the converse was true: Electricity can decompose chemical compounds into their basic elements. He applied electricity to alkaline earth and extracted magnesium, calcium, boron and strontium. In 1809, he demonstrated another use of electricity: the arc light. Before an audience, he held two thin charcoal sticks – one connected to a voltaic pile. After electricity began to flow from the battery to the first stick, Davey touched it to the top of the second stick and produced a dazzling spark. As Davey separated the charcoal sticks, the spark grew larger until he had created an arc of light.

Inventors tried to turn Davey's demonstration into a commercial arc light. Devising a sufficient power source stymied these efforts. A battery was not yet available that could economically run arc lights for a useful amount of time.

From Static to Dynamic

In 1820, Hans Christian Ørsted, a physics professor at the University of Copenhagen, gave a lecture about electricity. During a demonstration, he held a wire charged from a voltaic pile. He had intended to show how an electric current heats a platinum wire, but he became distracted by a swinging needle of a nearby compass. The compass needle reacted as if the charged wire produced a magnetic force. Later, he discovered that the needle would swing into a position at right angles to a charged wire. The needle deflected in the opposite direction when Ørsted reversed the flow of electricity. Ørsted had discovered that electric current induces a magnetic field.

Parisian professor of mathematics André-Marie Ampère was skeptical when he read Ørsted's report. Nevertheless, Ampère repeated the experiment. He proved that the strength of the magnetic field intensified with the rise of the power of the electric current. He also showed that wires charged with currents flowing parallel in the same direction attract each other, whereas wires with currents flowing in opposite directions repel one another. He made an electromagnet by running an electric current through coiled wire. The strength of an electromagnet increased as he added wire coils. Ampère created a stronger magnet by wrapping the coils around a piece of iron.

The work of Ørsted and Ampère came to the attention of Humphry Davey's assistant, Michael Faraday. If an electric current can generate a magnetic field, Faraday wondered, then can a magnetic field generate electricity? Faraday's experiments revealed that a change in magnetism, not steady magnetism, induced an electric current in a wire. In an early model, he wrapped an iron rod with wire and then moved a pair of strong magnets along it, producing current in the wire coil. Another version of Faraday's device had a copper disk placed between the poles of a stationary magnet. By rotating the disk, Faraday generated electricity, which could be conducted to a wire. Faraday had invented the electromagnetic dynamo. News of the discovery spread throughout the world. It was now possible to convert mechanical energy into electrical energy.

Faraday also advanced knowledge about electrochemistry. He created a new vocabulary to describe electrical phenomena, including the terms anode, cathode, ion, ionization, electrode, electrolyte, and electrolysis. Between 1839 and 1855, Faraday published his findings as a three volume set entitled *Experimental Researches in Electricity*.

At the same time that Faraday discovered induction, Joseph Henry of New York independently discovered the principle. Henry started his investigations into electromagnets around 1829. By insulating wire, he found that he could wrap more wire around a magnetic core without shorting out the device. His improved electromagnet lifted more than 2,000 pounds. Henry also invented an electric relay for a signaling device that could operate at great distances; this invention would develop into the electric telegraph.

Faraday and Henry had invented the dynamo, which converted mechanical motion into an electric current that coursed through a wire. Unlike batteries, dynamos could be constructed in ever-increasing sizes. The dynamo eliminated a barrier to the provision of electricity.

In 1832, Antoine-Hippolyte Pixii of France invented a machine considered to be the first workable electric generator. However, a practical application awaited the invention of a working dynamo by the Belgian engineer Zénobe-Theophile Gramme. By the early 1870s, Gramme had designed a powerful generator that produced current to mimic the steady flow of direct current produced by batteries. Gramme also designed a dynamo to generate an alternating electric current, in which the current periodically reverses direction.

The dynamo galvanized scientists and inventors around the world to devise practical applications for electricity. The generation of light offered one tempting target. Gas lamps had replaced oil and candles. Now, inventors wanted to replace gas lamps with a version of Davey's arc light. For a time, arc light could not compete with gas lighting; battery power proved too expensive.

In 1876, Russian military engineer Paul Jablochkoff devised a commercial version of an arc light, which generated a softer illumination than the early arc lights. The Jablochkoff candle paired two tall, thin carbon sticks separated by a layer of kaolin cement, which served as insulator and binder. The light had clusters of candles arranged so that when one burned out another automatically started. It could run for as long as 16 hours.

Despite improvements, electric arc lights were still more difficult to service than gaslights, and their dazzling glare restricted use to large spaces. By 1878, arc lights illuminated the elegant avenue de l'Opéra and other major Paris venues. Not everyone appreciated the change from gas light to arc light. "A new sort of urban star now shines out nightly, horrible, unearthly, obnoxious to the human eye; a lamp for a nightmare!" wrote Robert Louis Stevenson in *Virginibus Puerisque* (1881). "Such a light as this should shine only on murders and public crime, or along the corridors of lunatic asylums, a horror to heighten horror."

During the 1870s, arc light companies arose in America. In 1878, Thomas Edison examined William Wallace and Moses G. Farmer's "telemachon," a steam-run eight horsepower electric dynamo that powered eight arc lights. Edison told Wallace that the man had been working in the wrong direction; Edison could beat Wallace and Farmer in the electric light business. By this time, the arc light trade had captured ten percent of the street gas lighting business. Edison was more ambitious. He wanted to bring electric lights into homes.

Electricity Powers Society

A light bulb sparked the development of electric power systems in the United States. On December 3, 1879, Thomas Edison demonstrated his incandescent light bulb, an invention built upon the work of many scientists dating to the 1820s. To sell his light bulb, Edison would have to deliver an electric current to power it. In September 1882, Edison founded the world's first electric utility in lower Manhattan, a coal-fired company that supplied electricity to 59 paying customers. Edison's company generated direct current, which had its drawbacks. Since direct current transmitted at a single voltage, the current would burn out any appliance designed to run on less electricity, while an appliance would not work if it required more electricity. Another problem was that direct current could be transmitted over wires for only about one half mile. Edison's vision of an electric utility system would require coal-burning power stations to crowd the landscape.

In the spring of 1885, George Westinghouse, a Pittsburgh inventor and entrepreneur, read about an alternating current system on display at London's Inventions Exhibition. The system used a secondary generator, or transformer, to step down high voltages to those low enough to power an incandescent light bulb. Westinghouse thought that he could transmit high-voltage alternating current over vast distances and then use a transformer to step down the high voltage before it entered a house, factory, or office building.

Westinghouse sent an agent to Europe to investigate the alternating current transformer and obtained the rights in the invention. He then rebuilt the transformer and tested it. A month later, he incorporated the Westinghouse Electric Company. In March 1886, Westinghouse set up America's first centralized alternating current grid in Great Barrington, Massachusetts. By 1888, alternating current grids were set up or under construction in dozens of cities, competing with the direct current grids.

Edison became furious when he learned that Westinghouse dared to invade his electrical domain. In 1888, he published *A Warning From the Edison Electric Light Company.* Adorned with a blood-red cover, the booklet attacked Westinghouse and characterized alternating current as dangerous and destructive. "High pressure [i.e., high voltage], particularly if accompanied by rapid alternations, is not destined to assume any permanent position," Edison predicted. "It would be legislated out of existence in a very brief period even if it did not previously die a natural death."

Meanwhile, Nikola Tesla, a former Edison employee, had been creating a series of inventions based on alternating current. One invention would solve a problem for alternating current grids. By now, direct current motors were developed to power machines that had used steam, such as lathes and mills. Westinghouse needed an efficient motor powered by alternating current. Tesla created one. During the summer of 1888, Westinghouse acquired the rights to Tesla's alternating current induction motor, which enabled alternating current to power factory machinery and household appliances.

Westinghouse vigorously pursued the electric power business. He underbid Edison to supply power to the 1893 World's Columbian Exposition in Chicago. In 1896, Westinghouse founded the world's first hydroelectric generator at Niagara Falls, which transmitted electricity 25 miles to Buffalo, New York.

Edison sold his patent rights and generating stations to a company formed by financier J. Pierpont Morgan. It signaled the end of direct current utility companies. The new General Electric soon adopted the alternating current system for generating and transmitting electricity.

Samuel Clemens helps with an experiment during a visit to the laboratory of Nikola Tesla. Source: Martin, Thomas Commerford, "Tesla's Oscillator and other Inventions," *The Century* (April 1895).

II. Food

A Bird's Eye View of Frozen Food

A Scotsman introduced Great Britain to the Chinese practice of using cold to preserve food. In the 1780s, Alexander Dalrymple, stationed in China while working for the British East India Company, noticed that fishermen carried ice on their boats for fish caught at sea. He reported his observations to George Dempster, third Laird of Dunnichen and Member of Parliament. Dalrymple had good timing. On a tour of the Scottish Islands, Dempster realized that taxes on curing salt kept fisherman and their families in poverty. After reading Dalrymple's report about how Chinese fishermen preserved their catch with ice, Dempster asked a salmon dealer named Richardson to give it a try. The merchant packed freshly caught salmon in crushed ice and sent the fish to London. Six days later, the fish arrived in great shape. Before long, Richardson and other merchants routinely sent shipments of salmon packed in ice.

By 1814, ice had become the primary means to preserve fresh fish in Scotland. The British fishing fleets eventually carried ice to sea. The growing demand raised prices for purchasing and storing natural ice, which in turn, boosted interest in alternative cooling methods. Artificial refrigeration made a timely appearance.

In 1876, French engineer Charles Tellier decided that he could ship a leg of mutton across the Atlantic, and that it would arrive in perfect condition. Tellier fitted his steamer, re-named the *SS Frigorifique*, with his vapor-compression refrigerating engines and insulated cold rooms. The ship sailed from Rouen, France and arrived 97 days later at Buenos Aires, Argentina. Although the meat had not spoiled, it lacked an enjoyable flavor.

Cold storage technology soon improved. In 1881, the *London Times* published a letter from William F. Cott of the Bell-Coleman Mechanical

Refrigeration Company, who announced the arrival of about 5,000 carcasses of New Zealand mutton. "This shipment differs from all other importations of frozen meat," he wrote, "from the fact of having been made in a sailing vessel, which has been 98 days on the passage, during which time the holds of the ship containing the meat have been kept at about 20 degrees below freezing point."

Exporting frozen meat and fish became commonplace by the beginning of the 20th century. While popular with shippers, consumers gave frozen foods a chilly reception. Some of this reluctance might have been due to a belief that frozen meat could cause food poisoning. The major reason for consumers' lack of enthusiasm, however, resided in the destructive effects of freezing methods. Food frozen at the customary slow rate suffered from the formation of large ice crystals that ruptured the cellular structure of meat and other fare, destroying texture, flavor and natural juices, as well as reducing vitamin content.

Clarence Birdseye devised a way to produce frozen food that would be both palatable and healthful. Birdseye, who had majored in biology during his two-year college career, had worked for the US Department of Agriculture's Biological Survey as a naturalist. During this time, he traveled extensively and acquired a reputation for his willingness to eat unusual food, including the meat of mice, gophers, chipmunks, packrats, skunks, rattlesnakes and polar bears.

In 1916, Clarence Birdseye, his wife Eleanor, and their young son moved to Labrador, where Birdseye traded in furs. They lived in a three-room cabin located about 250 miles from the nearest store. Preserving food posed little difficulty. Clarence and Eleanor kept fish and game outdoors, frozen by the Arctic winds. When vegetables arrived in the area, Birdseye placed them in barrels of salt water where they quickly froze.

Perhaps due to his experience with unusual types of meat, Birdseye noticed an important feature about frozen food. Fish and game would be more flavorful if frozen quickly in the winter than in the spring or autumn. He also had the opportunity to observe the Inuit quick-freeze method, which preserved fish's flavor and texture.

Clarence and Eleanor started experimenting with freezing techniques. These experiments confirmed that quick-freezing avoided problems associated with commercial approaches for slower chilling. Quick freezing reduced the amount of large ice crystals, allowing thawed food to retain its original flavor, texture and appearance. The challenge would be to duplicate the quick freeze process in an environment that lacked Arctic cold.

The Birdseye family returned to the United States after the country entered the First World War. During the early 1920s, Birdseye resumed his experiments with quick freezing. Sequestered in a corner of the Clothel Refrigerating Company, he used an electric fan, brine and dry ice to cool fish to the freezing point by circulating cold air. Soon, newly-formed Birdseye Seafoods Inc. offered chilled fish fillets. Unfortunately, his venture was ahead of its time. Few grocery stores had freezer cases and consumers had no way to store frozen fish in their homes. Birdseye's business went bankrupt.

In spite of the setback, he continued to work on his freezing process and invented a method for quick freezing dressed fish or other processed food in cartons. To promote his flash-freezing invention, Birdseye founded General Seafood Corporation. In 1925, he moved his company to Gloucester, Massachusetts. Here, he developed a quick-freeze double-belt machine, which consisted of two long belts of stainless steel, one slightly above the other. Cold brine was sprayed beneath the lower belt and above the upper belt. Packages of processed food moved between the belts and froze immediately. The original machine weighed an unwieldy 20 tons. Smaller, multiple-plate versions of the quick freezer eventually became widely used in the frozen food industry.

Marjorie Merriweather Post, daughter of the Postum Company's founder, happened to be sailing on her yacht in 1926 when the ship docked at Gloucester harbor for provisions. While there, she ate a roast goose that was so tender and had such a superior flavor that she asked the chef where he had obtained it. The chef told her that he had bought it from an eccentric who experimented with freezing food. He also mentioned that the goose had been frozen for several months.

Post became determined to persuade the Postum Company's board of directors to buy Birdseye's struggling business. Three years later, she succeeded. Postum paid Birdseye 22 million dollars and changed its name to General Foods Corporation. Birdseye remained associated with the company as the head of the research and development department of the Birds Eye Frosted Foods division.

On March 6, 1930, the company performed a test marketing of its Birds Eye frozen foods in 18 stores in Springfield, Massachusetts. The trial run offered 26 items, including 18 cuts of frozen meat, spinach and peas, a variety of fruits and berries, oysters and fish fillets. Despite the success of this test and several others, obstacles remained before the frozen food industry could flourish.

By 1933, only about 500 grocery stores had freezer space for selling frozen foods. And, once again, consumers tended to be wary of these products. Birdseye's original business had licensed patents on its processes to a number of companies that manufactured frozen foods under their own labels. Many of the products were of poor quality. To address these problems, General Foods repurchased the licenses to control the quality of frozen foods, and leased freezer display cases to grocers at a low cost.

The US frozen food industry thawed during the Second World War. At the beginning of the country's involvement, the military placed orders for 70 million pounds of frozen foods, a windfall that financed improvements in food processing and production capacity. In 1944, at a time when fresh and canned foods were in short supply, the government lifted rationing on frozen foods. This bolstered widespread acceptance by consumers. During the next ten years, the introduction of the home freezer and the increased availability of frozen, precooked foods ensured the success of the frozen food industry.

Boston's Great Molasses Flood

Around 12:30 PM on January 15, 1919, soldiers newly returned from war heard a noise like machine gun fire, a sound they did not expect to hear in the city of Boston, a sound created by the popping rivets of a bursting tank that held more than two million gallons of crude molasses. With a roar, the tank exploded. A hot geyser of yellow-brown molasses propelled the top of the tank into the sky. Two huge sections of the exploding tank flew in opposite directions. A one-ton piece of steel severed a column in a trestle of elevated railroad tracks.

A molasses tsunami hit the city – 14,000 tons of thick, warm molasses. A wave estimated between 15 and 30 feet high slammed into buildings at 35 miles per hour. The mass of hissing molasses ripped a wooden firehouse from its foundation, demolished homes, and smothered men, women, children, horses and household pets. The freak disaster killed 21 people and injured 150.

The liquid cooled, leaving several blocks of downtown Boston buried in two to three feet of hardened molasses. Rescue workers trudged through the molasses-coated streets in their search for survivors. When they found living victims, the rescuers had to cut the stiff clothes away before rendering medical aid. Thick molasses coated the dead like hardened lava.

Several circumstances set the stage for Boston's Great Molasses Flood. The storage tank had been built in 1915, during a peak in the demand for molasses, which companies converted to industrial alcohol for military ammunition in the First World War. After the molasses disaster, an engineer concluded that the hastily-built tank had a faulty design.

The end of the war in November 1918 brought an end to the tremendous need for industrial alcohol. Fear of an impending Prohibition spurred companies to convert molasses into rum as quickly as possible before the ban. The Boston storage tank became filled to near capacity with crude molasses.

Then, the temperature rose. Between January 12 and January 14, the temperature soared from two degrees Fahrenheit above zero to 43 degrees. The molasses started to ferment. Gas strained against the tank's weak walls. The pressure transformed the tank into a time bomb, one that stood at the edge of the city's most densely populated neighborhood.

Scene of the tank explosion. Source: *The Boston Daily Globe*, January 16, 1919.

The Smith Butchering Machine

One story about Edmund A. Smith, possibly apocryphal, takes place around the end of the 19th century when he ran a cookhouse in Cascade, British Columbia. Business had turned sluggish, the story goes, and during his idle hours, Smith concocted a race to be held at an upcoming community sports day. The winner would earn a cash prize. After donating the first $50, he invited the town's residents to join him in subscribing for prizes and accumulated $500. Smith drafted the rules for the race, and apparently, no one noticed that the rules designated a $300 prize for an event called the "Fat-man's race." Smith happened to be a very large man.

On the day of the race, Smith gamely ran down the main street with the other contestants. He brought up the rear, gasping "I win! I win!" as he heaved himself over the finish line. The crowd laughed at Smith until he pointed out the fine print in the rules: A contestant had to weigh more than 280 pounds to qualify for a prize in the fat man's race. Smith weighed 320 pounds and was the only qualified contestant. Edmund Smith reportedly collected the prize money, but barely escaped with his life.

Soon, the large cook focused his creativity on inventing a machine that would revolutionize the Pacific Northwest's salmon canning industry. This time, his ingenuity would not be buried in fine print.

Mr. Smith Goes to Washington

Born on March 17, 1870 in London, Ontario, Edmund Augustine Smith spent his early years on his family's farm. As a boy, he designed and built a small threshing machine; it was a crude device, but it worked. After moving with his family to Victoria, B.C., Smith left home at an early age with little formal education. He tried various occupations, including cook and terra cotta presser and moved through the mining camps in Western Canada. In Cascade City, he met and married Gertrude Peterson.

On a trip to the Seattle region, Smith discovered a valuable clay deposit on a farm in an area that would become the town of Harper, near Port Orchard. In 1898, he settled in Colby and started the Harper Brick & Tile Co. with E.L. Grondahl, F.C. Harper, and Richard A. Ballinger, who became Mayor of Seattle and Secretary of the Interior during the Taft Administration.

Smith's surroundings must have aroused his predisposition to invent. After observing the deterioration of wooden pilings, he conceived the notion of pilings with a core surrounded by a layer of pottery clay "as the pile is thus rendered proof against the ravages of toredos or like subaqueous worms and the corrosive effects of salt water." His patent application for a "composite pile," drafted by the end of 1900, explains that the piling core may be made of cement and metal. This, at a time when the use of reinforced concrete for buildings and other purposes was unknown. Although Smith obtained a patent, he lacked the money to follow through on the idea.

Edmund Smith eventually sold his interest in the brickyard at a profit, moved to Seattle, and invested in the stock of the Alaska Fishermen's Union. The organization operated a cannery along the Chilkat River in Alaska. John Wallace and Benjamin R. Brierly shared an interest in the cannery, and the three became friends and eventually, business associates. His composite piling invention might have been ahead of its time, but in 1901 he turned his efforts to another invention that could not have been timelier. And for this project, he found enthusiastic backing.

According to an account in the *Pacific Fisherman*, a publication edited by Daniel L. Pratt, a close friend of Smith, a frivolous whim led to Edmund's greatest invention. Apparently, Smith was lounging in the Seattle office of the Alaska Fishermen's Union when a boy dropped by and tried to sell pencils personalized with an organization's name. Smith learned that the young salesman sent his orders to an East Coast company, which would produce the labeled pencils. The boy claimed that he could buy blank pencils at a very low price and make a lot of money if he had a pencil-printing machine. Smith told him that he would make the contraption for a hundred dollars. When the boy said that he did not have the money, Smith proposed that he would construct the machine, print the pencils and give the printer to the boy in exchange for the first $100 dollars. In less than one hour, Smith constructed a simple printing device that included a roller with rubber type. With Smith printing the pencils, the boy made $100 in a few days and walked out of the Union office with his little printer. Smith later claimed that he could have become rich by selling the machine at $1.50, but that the bottom of the market dropped out after the first sale.

F.E. Barlow, superintendent of the Chilkat cannery, did not view Smith's brief pencil-printing enterprise with amusement.

"Smith, why don't you turn your inventive genius to some practical use?" Barlow said. "Why don't you invent something that will do yourself and others some good?"

"Just name it," said Smith, "and watch me get on the job."

Barlow explained that his cannery was losing money for stockholders like Smith, because there was a bottleneck at the fish butchering tables. The company lacked sufficient labor to clean the fish fast enough to supply the lines of canning machines.

"Why don't you get up a machine that will clean fish in the canneries?" Barlow asked. "There have been a hundred or so invented and none of them are any good. The man that gets up a good one will make a fortune."

Barlow did not exaggerate the potential value of a fish-cleaning machine. The salmon canning industry was in dire need of automation at the butchering table, a circumstance created by the country's treatment of Chinese immigrants.

Intolerance Endangers the Salmon Canning Industry

Between 1849 and 1877, two hundred thousand Chinese citizens, ninety percent men, arrived in the United States. While most went to California, a substantial number settled in the Pacific Northwest. The California Gold Rush brought the first wave from China. Then the Central Pacific Railroad had to be built to meet the Union Pacific. In 1867, the railroad encouraged the immigration of additional laborers, and as one historian wrote, Chinese men "had to be brought across the Pacific, often without being consulted."

After completion of the railroad, the Central Pacific laid off most of the Chinese laborers, throwing thousands out of work.

The railroad's backers anticipated that the transcontinental railroad would bring prosperity to California. Instead, it brought inexpensive manufactured goods that hurt local industries. The railroad also brought unemployed European immigrants from the East Coast, who joined thousands of ex-miners, discharged railroad laborers, and former Union and Confederate soldiers, all seeking work. California's economy joined a national economic depression. Violence broke out in California against the Chinese, who had become the scapegoat for the poor economy.

Congress attempted a solution of sorts: the Chinese Exclusion Act of 1882. This legislation established a ten-year moratorium on the immigration of Chinese citizens, except for certain select groups, such as diplomats and their servants. The 1892 Geary Act extended the Exclusion Act, and ten years later, Congress decided to maintain the restrictions for an indefinite period of time. The Geary Act regulated Chinese immigration until the 1920s.

The Exclusion Act reduced the number of Chinese immigrants from more than 8,000 in 1883 to 10 in 1887. However, the legislation could not prevent the rise of another anti-Chinese movement, this time in the Pacific Northwest.

During the mid-1880s, Washington Territory experienced a sharp economic downturn in the wake of a sawmill curtailment and completion of several railroads, including the Canadian Pacific Railroad. Violence against Chinese immigrants broke out in Issaquah, and anti-Chinese forces expelled Chinese residents from Tacoma and Seattle.

While restrictions on immigration decreased the number of Chinese entering the United States, those in the country were leaving to avoid persecution. The decline in the Chinese work force impacted the Pacific Northwest's rapidly growing salmon canning industry, an industry that relied almost exclusively upon Chinese men to butcher the fish.

The butchers were the most skilled of the salmon canning crew. A good butcher could remove fins, head, tail and entrails with eight knife strokes and dress up to 2,000 salmon in one ten hour day. After the fish were butchered, they were sent to the "slimers," who scraped the fish to remove the mucous covering, some scales, and any blood or offal. Then the salmon were cut into small pieces and fit into cans that other workers had salted. The speed of the entire canning process depended upon the pace set by the butchers. And speed became critical during a run.

To spawn, salmon travel from the Pacific Ocean every summer into Puget Sound and from the Sound to various freshwater streams and rivers. The five species of Pacific salmon migrate at particular times from spring through fall. For example, King salmon is the first to arrive each year and migrates in early spring. This yearly migration, or "run," might continue for only a few weeks, so canneries made the best of it.

The experience of Pacific American Fisheries during the peak of the 1900 sockeye salmon migration illustrates why canneries employed thousands of workers to process the fish during a run. On August 1, the company's Fairhaven cannery received 85,000 salmon. From August 3 to 6, the cannery received a total of 232,000 salmon. On August 7, it received 70,000 salmon in the morning and 40,000 in the afternoon. It's no wonder that canneries had to operate almost continuously during the busy parts of the season.

Yet at a time when the canned salmon industry was rapidly growing, the availability of Chinese salmon butchers was diminishing. An account in the June 1909 issue of the *Pacific Fisherman* explained that:

> The Chinese laborers were skilled and difficult to replace. The training of new men meant the loss of much time and money. Moreover, it was impossible to get other laborers who were willing to do this work. The situation was a serious one and might have resulted in inestimable damage to one of the greatest industries in the West.

The canneries faced an additional problem: At the time of the salmon canning industry's peak development, the majority of Chinese men who remained in the business were getting old. The exhausting labor of a salmon butcher would begin hours earlier in the day than the rest of the crew, because they had to butcher a quantity of fish before packing could start. During the extended shift, salmon butchers worked with long sharp knives, their hands continually in water and fish guts. The butchers' feet and ankles became swollen from hours of standing in water and fish slime. Salmon butchering was not a job for the elderly.

The function of salmon butcher was ripe for automation. But the creation of a fish-butchering machine presented several obstacles. The process was complicated and needed precision to waste as little salmon as possible. By 1901, over 250 fish-cleaning devices had been patented and cannery men had tried many in their plants. None were successful.

A Revolution on a Dare

Edmund Smith had never seen the inside of a salmon canning company at the time that Barlow dared him to invent a fish-cleaning machine. Nevertheless, Smith took up the challenge. After Smith agreed to devise an automatic fish butcher, F.E. Barlow and John Wallace took him to a cannery on the Seattle waterfront, where he watched Chinese butchers cleaning fish by hand. Noting that the workers grabbed the fish by the tail and cut with the knife toward the head, Smith resolved that his machine would imitate the technique as closely as possible.

Wallace and Barlow advanced Smith's living expenses and set him up in a ten by fourteen foot workroom in a building on the corner of First Avenue and Seneca Street. Several weeks later, Benjamin R. Brierly bought a substantial interest in the project. Smith began building prototypes with just a hammer, chisel and hacksaw, but soon managed to convince his partners to purchase a $35 turning lathe.

The inventor labored at his machine in the downtown workshop, and drew blueprints on tablecloths at home. After eight months, however, he found that he had only created a substantial debt.

During a 1969 interview with Mrs. Helen Smith Sallee, Edmund Smith's daughter, she described one night when her father returned home and told his wife Gertrude that he would give up and get a job to repay the investment money. Gertrude simply advised him that if he took a bath, he would feel better about the situation.

Her advice must have helped. Smith woke at 3 AM that night and exclaimed "Gert! I've got it."

Unable to hire transportation at that hour, Smith ran from Yessler Way and 16th Avenue to his workshop. He worked for ten solid days. "We didn't see hide nor hair of him," said Mrs. Sallee. "Then he came home, all smiles, and got dolled up. He went to the bank to borrow some more money and took a patent attorney to Washington, D.C."

Smith built his first fish-cleaning machine during the winter of 1901 – 02 and filed his first patent application in May 1902. The automatic butcher consisted of a simple framework supporting a cam-driven plunger that carried the fish on a horizontal plane past knives and cleaning devices. The machine was not elegant, but it cleaned fish.

Smith and his partners established the Smith Manufacturing Company in 1902, and Smith continued to perfect his machine. He also found time to file a patent application for a machine that weighs and assorts packages, such as cans filled with salmon.

After his first study of a salmon butcher's technique, Smith did not see the inside of a cannery until the fall of 1903 when he installed a version of the machine known as "Jumbo" in the Fairhaven (now, South Bellingham) plant of the United Fish & Packing Company, operated by E.B. Dudden. Edmund designed Jumbo as a vertical wheel that carried salmon past knives and cleaning attachments. The vertical orientation required significantly less floor space than Smith's prototype and conventional models that operated along a horizontal plane. Despite the innovative design, cannery men remained skeptical about Jumbo in light of the failure of 50 fish cleaning machines invented by others. Yet on the first day, Smith's machine cleaned 22,000 fish in nine hours, or about 40 fish per minute.

On December 1, 1903, Smith, Wallace, Brierly and Barlow incorporated the Smith Cannery Machines Company; John Wallace was named as president of the company. During the following year, Smith developed a smaller model of his machine and leased it to six canneries for a royalty: three in Alaska and three in the Puget Sound area.

Although the fish-cleaning machines operated successfully, Smith's business did not. Smith Cannery Machines Company never received a royalty payment. By the end of 1904, the company had yet to take in one cent.

Then, Everett B. Deming of Pacific American Fisheries, Inc. (Bellingham, WA) bought three machines in 1905, the first sale made by Smith Cannery Machines. Deming reportedly made the payment with Check No.1 of the newly incorporated cannery. Two automatic butchers supplied seven lines of canning machinery, which packed an average of 9,000 cases of sockeye salmon a day, and more than 10,000 on some days. For comparison, the company had operated in 1901 with nine canning lines and a large butchering crew working continuously to pack 8,600 cases on the best day. Smith boasted that the "iron chink kept them continually supplied and the lines of machinery never were idle for want of fish and frequently there were from 30,000 to 70,000 fish cleaned ahead."

Edmund Smith's comment highlights an infamous aspect of the fish-cleaning machine. Soon after it was put into operation, the device became widely known as the "Iron Chink," a derogatory term based on the claim that it performed the work of about 50 Chinese salmon butchers. A myth arose that Edmund Smith had developed the machine specifically to displace Chinese workers. This fiction might have been fabricated to stimulate interest among cannery owners who wanted to find a substitute for the relatively well-paid salmon butchers.

The Pacific American Fisheries sale probably saved Smith's company. Moreover, the cannery's success with the machine generated much-needed publicity. In 1906, eight new Smith Butchering Machines were installed in the United States and five in British Columbia. To keep pace with the increased business, Smith's manufacturing plant expanded its Seneca Street facility.

An article published in the May 1906 issue of *Pacific Fisherman* described the operation of the machine at that time.

The method of cleaning the fish is simple in the extreme. Two men are required to prepare the fish before they enter the machine, one of whom seizes the fish as it comes down the elevator and guides it past a knife which cuts the head off. The other passes it by a knife which cuts off the tail, and then the fish is ready for the machine and is placed in the trough which feeds the cleaning cycle of the "Iron Chink." The fish comes through the cleaning trough tail first, the back fins coming in contact with the self-sharpening knife which trims off the large and small fins. In the trough an automatic feed works consistently with the six clamps on the wheel, which clamp the fish by the tail, carrying them up through a centering device which holds them firmly when the back clamps close on them. The remaining fins are removed in uniform manner by self-sharpening, self-adjusting knives at the top of the machine, and the fish pass on down to the splitting saw which splits the fish in the exact center.

Further on the fish come in contact with a rotary, grappling device which removes the entrails and stirs up the blood on the backbone, and the fish are then ready to be washed out with the aid of a stream of water and a rotary brush, after which they pass on to a point within a few inches of where they entered the machine.

The fish then traveled on a conveyor to the gang knives and on to the canning machinery.

A limitation of the machine was that, although it could be set to handle fish of different sizes, it could not adjust itself from one fish to the next. Certain canneries, such as Columbia River canneries, received fish with a wide variation in size, and the use of the butchering machine under these conditions would require frequent resetting or manual sorting of fish by size. This would work against any potential savings of time and labor. Nevertheless, companies continued to purchase the butchering machines and place them in canneries where the salmon runs and type of salmon justified their use.

The early models of the Smith Butchering Machine frequently needed repair. Smith learned about his invention's defects by living in a cannery, observing the machine's operation, improving it, and sleeping between repairs. His improvements are reflected in four patent applications that he filed between 1903 and 1909. The 1908 model proved so satisfactory that no major alteration was introduced during the next ten years.

By mid-1909, more than 60 butchering machines were used at canneries located in Puget Sound, British Columbia and Alaska. The demand was so great that Smith's company announced plans for a new three-story manufacturing facility at First Avenue and Stacey Street – the first reinforced concrete structure in Seattle. Edmund Smith did not live to see it.

1909
ALASKA YUKON PACIFIC EXPOSITION
THE "IRON CHINK"
WILL BE IN ACTUAL OPERATION DAILY
DURING THE EXPOSITION
CAPACITY 1 FISH EVERY SECOND 3000

Preparing for the 1909 Alaska-Yukon-Pacific Exposition. Edmund Smith appears at the right of the photo. Source: Museum of History and Industry (Seattle, WA).

Seattle's Alaska-Yukon-Pacific Exposition was set to open on June 1, 1909. Smith assembled an exhibit for the event that would show visitors how his machine cleaned salmon. On May 31, Edmund drove his sister, Mrs. J. Sutcliffe, to see the display on the grounds of the University of Washington. On the way, they entered a blind alley about one block north of the Latona Bridge (now, the site of the University Bridge). While backing out, the automobile ran into a rut and rocks perforated the rear gas tank, which exploded. Burning gasoline drenched the occupants of the car. Although pinned beneath the steering gear and blinded by flames, Smith managed to shove his sister over the side of the car to the ground. He did not follow her because he was afraid that he would fall on her and that his weight would cause more injury. Instead, he tried to work his way over the brakes to the other side.

Rescuers arrived, disentangled Smith from the car, and rushed the two to Pacific Hospital. Edmund Smith's doctor said that he would have to amputate several fingers from the inventor, and he requested Smith's relatives to donate skin for extensive grafts. Although Mrs. Sutcliffe was severely burned, her doctor did not anticipate a need for a skin graft.

A *Seattle Times* reporter visited Smith on the morning of June 1. "I guess you'll have some trouble to understand me, because this isn't much fun," Smith mumbled between blistered lips. "You may say however, that soon as I am able I want all of those who so kindly helped me at the scene of the accident to call and let me personally thank them." Smith said that he and his sister would "be able to see the fair long before it closes and have a good time with our friends." But Edmund Smith died unexpectedly at 5:45 on the morning of June 2.

Daniel L. Pratt, who had known Smith since he first started his experiments with the fish-cleaning machine, told a *Seattle Times* reporter that the country had lost a second Edison. Pratt hinted at inventions that Edmund had been devising in the experimental department of his factory.

Edmund Smith might not have had the chance to realize his full potential as an inventor, but by the age of 39 he had certainly left his mark. His automatic butchering machine revolutionized the canned salmon industry. After Smith's machine removed the holdup at the butchering table, new production bottlenecks arose, forcing the development of new machines and the improvement of older technology. As Patrick W. O'Bannon wrote in his 1982 *Agricultural History* review, the industry's adoption of the automatic butcher "unleashed a wave of innovative activity both inside the cannery and on the fishing grounds." A report in the January 1927 of the *Pacific Fisherman* offered another measure of the invention's impact.

Machine butchering in salmon canneries, as performed by the famous "Iron Chink," has been among the greatest forward steps in the development of this branch of the fisheries, and is one of the principal factors making possible the increase of production from about 3,000,000 cases in 1900 to over 10,000,000 in 1926 and record years of war times.

Smith's fish-cleaning machine not only enabled the growth of the salmon industry, but ultimately created more jobs than it eliminated. United States and Canadian salmon canneries still use the latest model of the "Iron Butcher." The Seattle-based business, Smith Berger Marine, Inc., traces its roots to Smith's old company and promises to continue the legacy of Edmund Smith.

Candy Corn

For many North Americans, the phrase "candy corn" evokes an image of a small, triangular confection with a white tip, orange center and yellow bottom. The popularity of this eccentric candy has endured for more than a century.

Food historians debate about the identity of the candy corn inventor. Some credit George Renninger, employed by the Wunderlee Candy Company of Philadelphia. In the early 1880s, Wunderlee did become the first company to produce the candy with a corn kernel shape. By the end of the 19th century, North Chicago-based Goelitz Confectionery Company started production of the candy and has been doing so since.

In an age predating automation, Goelitz relied upon backbreaking manual labor to transform raw materials into the deceptively simple looking treat. From March to November, men toiled in an overheated kitchen, cooking sugar, water, corn syrup and other ingredients in large kettles. Into this sweet slurry, they whipped fondant – a sugar paste – for smooth texture, and marshmallow to add a soft bite. When the mixture reached the desired consistency, they poured the hot candy into hand-held buckets, each holding about 50 pounds. Walking backward, workers poured steaming candy into trays of cornstarch imprinted with kernel-shaped molds. This required three passes: one for each of the yellow, orange and white portions. Creating the little candy took strength, dexterity and stamina.

The public rewarded these efforts; candy corn became a bestseller. The novel tri-color design amazed candy lovers. Farmers in particular appreciated the candy's agrarian appearance. Competitors took note and introduced candy in other vegetable shapes, including turnips, but none seized consumer's interest like candy corn. Sales of the candy carried Goelitz through two world wars, the Great Depression, minor depressions, recessions, and the soaring cost of raw sugar during the mid-1970's.

Today, computerized machinery performs most of the work required to produce candy corn. The Goelitz recipe remains the same even though the company has changed its name to the Jelly Belly Candy Company.

Now sold year-round, candy corn represents the signature Halloween treat. Other holidays have their unique versions of candy corn: chocolate and vanilla-flavored Indian Harvest Corn for Thanksgiving; red, green and white Reindeer Corn for Christmas; red, pink and white Cupid Corn for Valentine's Day; and pastel-colored Bunny Corn for Easter.

Companies generate the sweet kernels in astronomical numbers. According to the Virginia-based National Confectioners Association, candy corn manufacturers produce more than 35 million pounds – about nine billion kernels – per year, an amount sufficient to circle the moon nearly 21 times if laid end-to-end.

Potato Chips

Food historians credit two individuals with the invention of potato chips: Thomas Jefferson and a vexed chef named George Crum. Jefferson, who served as US minister to France, introduced US citizens to fried potatoes, cuisine that he discovered in Paris. Moon's Lake House, a resort in Saratoga Springs, New York, offered the popular side dish on its menu in 1853. That summer, a fussy customer dined at the Lake House and sent his fried potatoes back to the kitchen. The cook made the potatoes unacceptably thick, he complained.

Chef George Crum tried to appease the patron by cutting and frying a thinner batch. These, too, met with rejection. Frustrated by the finicky diner, Crum, who had a reputation for orneriness, made a special batch of over-salted fried potatoes so thin and crisp that they could not be eaten with fork. To Crum's surprise, the diner loved the crispy potatoes. Other customers requested the fried chips of potato, which soon appeared on the menu as a house specialty called Saratoga Chips.

As the fame of Saratoga Chips spread, other New England restaurants began to serve their versions of potato chips. For several decades, crunchy potatoes remained a side dish served mainly in Northeast restaurants. By the end of the 19th century, however, William Tappenden of Cleveland began to make chips in his kitchen and sell them to neighborhood grocery stores.

In the early 1900s, potato chip factories sprouted throughout the United States. Mike-sell's Potato Chips in Ohio, Leominster Potato Chip Company of Massachusetts, Alabama-based Magic City Food Products Company, Utz Quality Foods of Pennsylvania, and other companies marketed the novelty food. These operations relied upon manual labor to prepare kettle-cooked chips.

Three innovations radically altered the budding industry in the 1920s. When Bill and Sallie Utz started their company, Sallie used hand-operated equipment to produce about fifty pounds of potato chips per hour. The mechanical potato peeler eliminated the need to peel and slice potatoes manually. The continuous fryer rendered small-scale kettle cooking obsolete. After the Utzs installed an automatic potato chip cooker, they could produce 300 pounds per hour.

The most significant of the three innovations seems the simplest: a moisture-resistant, sealed potato chip bag. Traditionally, groceries stored chips in barrels or display cases. Clerks weighed chips for their customers and placed the snack in paper bags. This did not present an ideal distribution method for fried food.

Laura Scudder, who worked in her Monterey Park, California-based potato chip company, devised a plan in which employees took home sheets of waxed paper and hand-ironed them into bags. The following day, they hand-packed chips into the bags and sealed the tops with a warm iron. With chips secured in bags, Scudder could offer customers self-service and ship a fresh product farther. Later developments in cellophane and glassine packaging further expanded the potato chip market.

Although now well preserved, chips lacked wide appeal until Irish chip manufacturer Joe Murphy revolutionized the industry. In the 1950s, Murphy's Tayto company produced "Cheese and Onion" chips, the world's first flavored potato chips.

Other potato chip manufacturers enriched their products with natural and artificial additives, producing the wealth of flavors that tempt today's consumers: sour cream and onion, roast chicken, prawn cocktail, Thai spice, sweet Wasabi mustard, honey Dijon, cheddar with beer, jalapeño with tequila and lime, roasted red pepper with goat cheese, steak and Worcestershire, buffalo wing bleu cheese, and others. Flavor transformed potato chips into a seasoned competitor among snack foods, one that brings more than $6 billion per year in US retail sales alone.

Fig Newtons

In the early 1890s, inventor James Henry Mitchell created a machine that extruded dough filled with preserves or jam. After baking, the jam-packed cookie cables could be cut into individual cookies. He contacted the Kennedy Biscuit Works in Cambridgeport, Massachusetts to see if they would be interested in the machine. Mitchell wisely chose his target company; Kennedy Biscuit Works made and sold biscuits, cookies and preserves. The company could use Mitchell's invention.

Soon, Kennedy Biscuit Works introduced Newton, a cookie stuffed with fig, the company's bestselling jam. Why call a cookie "Newton"? Plant manager James Hazen had a habit of naming new products after surrounding neighborhoods and suburbs. The company sold the Brighton, the Melrose, and Cambridge Salts crackers. For the new fig cookie, Hazen picked the Boston suburb of Newton.

The National Biscuit Company acquired Kennedy Biscuit Works in 1898. The company added "Fig" to the name of its most popular acquired product.

Nabisco marked the centennial of the Fig Newton's creation by releasing peculiar facts about the little cookie. A person would need 1,520 cookies to match the height of Niagara Falls. Acquiring 4,444 cookies would allow a Fig Newton fan to equal the height of the Washington Monument.

Cracker Jack®

Soon after the Great Chicago Fire of 1871, German immigrant Frederick William Rueckheim opened a popcorn stand in that city. His business expanded. In two years, his brother Louis helped him sell peanuts, caramels, chocolates, marshmallows and molasses taffy.

Frederick Rueckheim decided to combine three popular snack flavors: popcorn, peanuts and molasses. Snack eaters in the Northeast had enjoyed molasses-covered popcorn balls for 20 years. But the addition of peanuts, a novelty circus snack, created something unique. Rueckheim and his brother introduced "candied popcorn and peanuts" to the Chicago public.

During the following three years, Louis discovered a way to prevent molasses-covered popcorn from sticking. The brothers' snack transformed from a traditional ball shape to small nuggets. According to legend, a salesman tried the snack, loved the taste, and enthusiastically exclaimed, "That's a crackerjack!" Rueckheim agreed and secured rights to the phrase.

Until 1899, Cracker Jack had been sold in large tubs. Henry Eckstein, who had recently become a partner with the Rueckheim brothers, invented a waxed-sealed package to retain freshness. Boxed Cracker Jack could now be sold nationwide. By 1902, the snack had become so well known that the Sears catalog listed Cracker Jack without a description.

In 1912, Cracker Jack boxes included a new, vital ingredient: trinkets. These small toys further boosted the snack's popularity.

The First World War inspired Rueckheim Bros. & Eckstein to add another important element to the Crack Jack box. Amidst a patriotic red, white and blue color scheme stood a sailor boy named Jack and his dog. The pair had been modeled after F.W. Rueckheim's grandson Robert and his dog, Bingo.

Animal Crackers

During the late 1880s, Phineas Taylor Barnum decided to bring his circus to England, a land where children ate animal biscuits. Barnum's arrival inspired companies to sell animal biscuits wrapped in packages with circus designs.

Before long, many small American bakeries made animal or circus cookies, sold in wooden boxes and barrels. The popularity of animal cookies rose sky-high when the National Biscuit Company launched "Barnum's Animals" in 1902. Designed as a Christmas season novelty, the company sheltered the cookie beasts in red and green boxes decorated with a circus theme and animal pictures. Cookie lovers could hang the ornamental box by its string handle on a branch of a Christmas tree.

The item proved so successful in the public arena that the company manufactured Barnum's Animals throughout the year. In 1948, the company rechristened its product "Barnum's Animal Crackers."

During the past century, Barnum's Animal Crackers have included more than 50 different animals. To celebrate the 100th anniversary of the cookies, Nabisco ran a poll, asking the public to decide whether the walrus, cobra, penguin or koala should be added to the 17-animal menagerie. The votes came in favor of the koala.

The cookies' popularity endures. Each year, Nabisco uses almost 3,000 miles of string for the small boxes.

Humble Pie

"When I was quite a young boy," said Charles Dickens' Uriah Heep, "I got to know what umbleness did, and I took to it. I ate umble pie with an appetite."

Today, a person eats humble pie by admitting an error under humiliating circumstances. Humble pie has served as part of the idiomatic phrase since the early 19th century. Before then, humble pie was served on a plate.

The word "numbles" or "noumbles" referred to animal viscera during 14th century England. While upper classes dined on venison, hunters, servants and children ate inferior deer viscera in a numble pie. In the 15th century, the name changed to umble pie, possibly due to a transformation of "a numble pie" to "an umble pie."

The British exported the delicacy to North America. Susannah Carter's *The Frugal Housewife, or, Complete Woman Cook* (1803), published in New York, explained how to concoct an "umble pie" in one breathless sentence: "Take the umblers of a buck, boil them, and chop them as small as meat for minced pies; put to them as much beef suet, eight apples, half a pound of sugar, a pound and a half of currants, a little salt, some mace, cloves, nutmeg, and a little pepper; then mix them together, and put it into a paste; add half a pint of sack, the juice of one lemon and orange, close the pie, and when it is baked serve it up."

Perhaps, the umble pie acquired its "h" due to the longstanding association with inferiority. This change would have gone unnoticed by the h-dropping Uriah Heep.

Coffee Makers

According to folklore, an Ethiopian goatherd named Kaldi, who lived around 300 AD, found his goats elatedly butting heads and cavorting on their hind legs. They would stop occasionally to chew green leaves and red berries of a tree that Kaldi had never seen before. After chewing a few berries, the energized goat herder merrily joined the dance.

Kaldi's discovery of the intoxicating properties of coffee beans sparked a quest for new ways to release the beans' storehouse of caffeine and aromatic volatile oils. By 575 AD, the Turks boiled water with coffee beans in an *ibrik*, a tall, long-handled copper or brass pot that tapered to an open top. For centuries, coffee lovers boiled whole roasted coffee beans – and later, ground, roasted beans – to perform a method of extraction known as decoction.

In 1711, the French began to make coffee using an alternative to decoction: infusion. A coffee infusion, an extraction performed at any temperature below boiling, can be achieved by three methods. Steeping, the first and simplest method, requires mixing hot water with ground coffee beans in a container. Infusion can also be achieved by trickling hot water through fine apertures in china or metal (percolation) or through a porous material like cloth or paper (filtration).

Although infusion improved coffee's taste, coffee grounds still found their way into cups and made the drink taste bitter. Inventors enthusiastically designed ways to overcome the problem. Some designed coffee pots that trapped floating grounds in narrow spouts and settled grounds in broad bottoms. Others placed the spout in the middle of the pot to evade floating and settled grounds while pouring.

Filters offered another way to ensnare grounds. In the late 18th century, French pharmacist R. Descroissilles made a metal pot with an upper or inner section pierced with small holes that served as a filter. Hot water passed by the grounds, through the holes, and into the pot's lower chamber. The "biggin" became a popular way to control grounds, but did not solve the problem of over-brewing.

The 19th century experienced a proliferation in coffee making innovations. Jean Baptiste de Belloy, the Archbishop of Paris, invented a drip coffee maker. In Germany, Dr. Romershausen devised a way for steam to push water through the coffee, an early form of a steam espresso machine. Madame Vassieux of Lyons developed the *hydropneumatique* pot, which used a partial vacuum to pull hot water through ground coffee and a filter. Scottish marine engineer Robert Napier later improved this device. Jacques-Augustin Gandais, a Parisian jewelry maker, invented the pumping percolator. The French coffee press, or *cafetière*, minimized brewing time and enabled a quick removal of coffee grounds with a filter attached to a push rod.

Several types of coffee makers used metal filters by the early 20th century. Linen or cloth that lined the metal filters shredded and often imparted a rotten taste. In 1907, Mrs. Melitta Benz of Germany began to experiment with different materials to place between the two chambers of a biggin-like coffee pot. When she tried circles of blotting paper, she found that the paper made a perfect filter and simplified the disposal of coffee grounds. Mrs. Benz transformed her simple discovery into a company that still flourishes.

By this time, the basic methods for brewing coffee had been extensively tested. The greatest innovations of the new century lay in harnessing electricity to drive time-honed extraction techniques. Thanks to the introduction of the automatic drip coffee maker in the 1970s, drip brewing has become North America's most popular method.

Whistling Teakettle

As a young boy in New York, Joseph Block observed his father invent a new type of pressurized potato cooker, one that whistled when the cooking cycle ended. Block stored away the memory and became a cookware executive. After retiring, he toured a teakettle factory in Westphalia, Germany. This was in 1921, a year that brought a revolution to tea drinking.

Block must have remembered his father's invention while he explored the German factory. What about a teakettle that whistles, he suggested. To test the idea, the factory produced 36 whistling teakettles. A local department store sold the entire lot within three hours.

In 1922, Block premiered his whistling teakettle at a housewares fair in Chicago. During a week of demonstrations, he kept one kettle whistling at all times. Wanamaker's department store in New York took 48 kettles. Store executives saw the one dollar whistling kettles draw customers into the housewares department. Before long, stores across America sold 35,000 whistling teakettles a month.

Whistling teakettles did not go unnoticed in Canada. The September 5, 1933 issue of Alberta's *Lethbridge Herald* highlighted the device in its "Of Interest to Women" column:

We were intrigued with the new whistling tea kettle that was a shower gift to the bride-elect the other day. Of light aluminum and made to heat quickly, it has a little whistle fastened to the spout, so that the hostess may put the kettle on and go to talk while it boils. A warning note tells her when it is time to brew the tea!

Tea Bags

The tea bag radically changed time-honored practices of tea drinking worldwide. Serendipity sparked this minor revolution.

In 1904, Thomas Sullivan, a New York tea and coffee merchant, concocted an advertising gimmick. He handed out samples of loose-leaf tea sealed in hand-sewn silk bags. Tea drinkers didn't bother to open the bags before brewing their tea. They used the bags like metal infusers and dropped the bags into hot water.

Sullivan must have been surprised to hear his customers complain about the fine mesh of the silk bag, which he had intended to serve only as an attractive container. Nevertheless, Sullivan responded to the complaints and devised packets made of gauze.

Tea drinkers were not the only ones who appreciated bagged tea. Restaurateurs discovered that tea bags eased the process of cleaning after their tea-drinking patrons.

Tea made from early tea bags tended to taste too much like the material used to make the container. Manufacturers tried an assortment of materials for their bags. Hemp, rayon and wood pulp failed to produce satisfactory results. Gauze made an acceptable bag. Later, a flavorless paper fiber became the dominant material. Companies marketed tea bags in America during the 1920s.

In 1953, Tetley introduced the tea bag to Britain. Initially wary, the British eventually appreciated the convenience of the tea bag. During the 1960s, tea bags usurped the traditional tea pots in Britain. America had exported its revolution in tea drinking, and perhaps, made amends for the Sons of Liberty, who tossed several hundred crates of tea into Boston Harbor centuries earlier.

Crisco®

In 1837, William Procter and James Gamble started to make and sell two byproducts of the animal processing industry: soap and candles. By 1870, Proctor & Gamble also sold lard, another byproduct.

The 20th century ushered new challenges for Proctor & Gamble. Electric power burned a hole in the candle market and meatpackers captured the lard market. The company needed a unique product.

E.C. Kayser, a German chemist, wrote to Proctor & Gamble officials in 1907. He wanted to know whether they would have any interest in his investigations of hydrogenation, the process of using hydrogen to convert liquid oil to solid fat. Proctor & Gamble's researchers had worked on a way to treat cottonseed oil so that it could remain liquid in cold temperatures. But they had not achieved the objective of creating a solid form of cottonseed oil that could compete against lard and butter. Kayser went to work for Proctor & Gamble.

In 1911, the company launched Crisco, the first solidified shortening product made entirely of vegetable oil. Crisco cost less than lard and butter and could be heated at higher temperatures without burning or smoking. Food absorbed less Crisco, compared with lard and butter, and this added to the product's economic benefits. Consumers were nonplussed.

Proctor & Gamble launched a massive campaign to educate consumers about their unique product. The company published books with Crisco recipes and new cooking techniques. A team of home economists traveled across the United States. They taught methods for using the vegetable shortening to consumers accustomed to traditional cooking fats.

Consumers eventually accepted Crisco through a combination of advertising efforts and shortages of lard and butter during the First World

War. Crisco became the best-selling household vegetable shortening in the United States and transformed the American diet.

The story that had to be told (1914 edition). A 1921 edition offered 615 recipes.

The Corkscrew

The corkscrew, a tool that supplies an irresistible force to remove a seemingly immovable object, evolved with wine storage technology. Before the availability of glass bottles, barrels, earthenware jars and animal skins stored wine. The wood or cloth stoppers that sealed these containers could not create an airtight seal. Consequently, wine could not be safely aged, because the plugs failed to prevent oxidation to vinegar.

Advances in glass-blowing expertise enabled the production of bulbous, onion-shaped bottles designed to stand upright on a shelf. A wood plug or tapered cork bound with waxed linen wrapping sealed the "shaft and globe" bottles and offered a convenient handhold for pulling out the stopper.

During the 17th century, further advances in glass fabrication brought bottles with more uniformly sized necks. The English may have been the first to seal the new wine bottles hermetically with cork imported from Spain and Portugal. Now stored in airtight bottles, wine improved with age, rather than spoiled.

As manufacturers produced bottles with cylindrical shapes and narrow necks, corks had to be compressed and softened before insertion. The challenge of removing the tight-fitting cork encouraged the invention of an instrument, possibly inspired by the metal screw of a tool used to clean gun barrels. In 1681 England, the cork extractor, or "bottlescrue," came to the aid of wine drinkers.

The first corkscrews were uncomplicated T-shaped devices consisting of a wood handle and a worm, a pointed and curled piece of metal. To open a bottle, a wine enthusiast turned the metal worm into the cork, and then used brute force to wrench the cork from the bottle.

Reverend Samuel Henshall of Birmingham, England, advanced the art of cork extraction with his 1795 patent, deemed the first patent awarded for a corkscrew. Henshall added a concave disk between the worm and the shank. After the worm was driven into the cork, the disk contacted the top of the stopper, forcing the cork to turn with the twisting of the crosspiece, which in turn, broke the adhesion that bound the cork and the bottle's neck. For over a century, England produced button corkscrews, which became known as "Henshalls." Some of these came equipped with brushes to wipe away cobwebs and dust from bottles of aged wine.

T-shaped corkscrews and raw force proved an adequate combination for a while. But the 19th century ushered improvements in bottle making and cork fitting, innovations that increased difficulties in cork removal.

Inventors eagerly confronted the latest challenge to accessing bottled wine. German Carl F.A. Wienke invented a corkscrew with a single lever that aided in lifting a cork. Known as the "Butler's Friend," or "Waiter's Friend," this design is still in use today. In 1862, American Charles Chinnock obtained a patent for his "self-puller" corkscrew. The device had a cylindrical frame attached to the shank of a worm. After the frame was placed on the neck of the bottle, a handle was turned to drive the worm into the cork. The cork conveniently rose on the worm. In 1888, H.S. Heeley received a British patent for a more familiar design: the double-winged corkscrew. To remove a cork with this device, the wings are lifted, the worm is twisted into the stopper, and then the wings are squeezed toward the bottle's neck, which raises the cork.

Corks and corkscrews served many purposes before bottle caps and cans became common after the Second World War. Corked bottles stored a variety of products, including beer, whiskey, ketchup, relish, ink, poison, medicine and perfume.

Toasters

The practice of toasting bread predates the toaster by thousands of years. The ancient Egyptians toasted bread to dry and preserve the food. Toasting also improves the taste of bread. As bread turns brown from caramelized sugars, it becomes sweeter.

Over the centuries, people have toasted bread on hot stones or over open fires with toasting forks. Indoors, hearth toasters proved a handy tool. Attached to a fireplace, these hinged frames held a piece of bread and could be swung toward the fire.

By the end of the 19th century, inventors had designed appliances to run with electricity. Toasters eluded this trend – with one exception. The British firm, Crompton & Company, introduced an electric toaster in 1893. But the appliance was ahead of its time; a filament had yet to be designed that heated quickly without burning out or starting fires.

In 1905, Albert Marsh patented NiChrome, a filament made of a nickel and chromium alloy, which had a high melting point. General Electric incorporated the filament in its 1909 toaster. Electric toasters became popular even though the bread had to be carefully watched and turned to heat both sides.

Charles Strite of Minneapolis introduced two innovations by 1920. First, he designed a toaster that heated bread for a fixed time and then turned off. Strite also invented a toaster with a timer and a spring, which popped up the toasted bread.

In 1926, the Waters-Genter Company launched Strite's appliance, the Toastmaster. Unlike many early toasters, the Toastmaster heated both sides of bread simultaneously. "You do not have to watch it," the company boasted, "The toast can't burn." The convenience of automatic toasting, coupled with acceptance of standardized, sliced bread, ensured that toasters would always have a place in the kitchen.

Food Mixers

In 1908, Herbert Johnson, an engineer with the Hobart Manufacturing Company (Troy, Ohio), watched a baker laboriously mix bread dough with an iron spoon. Johnson thought of a way to help bakers: He invented an 80-quart machine that mixed bread dough, an easier and more sanitary alternative.

Johnson's stand mixer turned out to be a great success. By 1917, it had become classified as standard equipment on all US Navy ships. Encouraged by the machine's acceptance, the company decided to produce a mixer for the general public. The First World War delayed the company's plans.

After the war, the Troy Metal Products Company, a subsidiary of Hobart Manufacturing, began production of the H-5, a stand mixer for the home. The machine's beaters mixed food with a planetary action. Like a planet moving around the sun, the beaters rotated in one direction while they moved around the bowl in the opposite direction.

According to company history, the wives of Troy's executives discussed a name for the new contraption. "I don't care what you call it," said one woman, "but I know it's the best kitchen aid I have ever had." In the 1920s, the company advertised how the KitchenAid® stand mixer can "stir, beat, cut, cream, slice, chop, and strain, by electricity!"

In response to reluctant merchants, the company sold the H-5 door-to-door. During the 1930s, a smaller, less-expensive mixer gained wide acceptance.

III. Health

Diabetes Treatment Comes of Age

According to the World Health Organization, more than 190 million people have diabetes, a number that may exceed 360 million by 2030. Diabetes occurs in two principal forms. About ten percent of diabetics have Type 1, which arises when the pancreas no longer produces insulin – the hormone that regulates carbohydrate metabolism. Type 2 diabetes, the more common form, is a group of disorders characterized by variable degrees of impaired pancreatic insulin secretion combined with an inability to use insulin. Whereas insulin treatment can be necessary to treat a person who has Type 2 diabetes, it's vital for the survival of a Type 1 diabetic.

The birth of medicinal insulin occurred at 2 AM on October 31, 1920. That is when Dr. Frederick Grant Banting envisioned a method for isolating an anti-diabetic compound from pancreas. His initial experiment required less than one month to complete. Yet it marked the end of several thousand years of failed treatments.

Early Perspectives on Diabetes

Classic symptoms of diabetes include excessive urination and high levels of glucose in the blood and urine. People have observed these signs for ages. An Egyptian papyrus, written around 1550 BC, describes the diabetic symptom of excessive urination. This information may have been copied from books written 1500 years earlier. In the 4th century BC, Hindu physicians Charaka, Susruta and Vaghbata noticed ants congregating around the sweet urine of people who had a condition that they called *Madhumeha* (honey-like urine). Four hundred years later, Arataeus of Cappadocia named the condition characterized by copious urine output *diabetes*, from the Greek word meaning siphon or pipe-like.

Chen Chhuan of the 7th century AD in China and Avicenna, an Arab physician of the 11th century, also described a condition associated with sweet urine. The observations of Oxford physician Thomas Willis in the 17th century and the experiments of Manchester physician Mathew Dobson during the 18th century established the diagnosis of diabetes as the presence of excess sugar in the urine and blood. William Cullen, a British clinician and contemporary of Dobson, designated this form of diabetes as *mellitus* – from the Latin word for honey-sweet.

Healers have experimented with various methods to treat diabetes. Around 1000 AD, Greek physicians prescribed exercise. Eight hundred years later, John Rollo, the English Surgeon General to the Royal Artillery, promoted a diet of rancid meat. A fasting therapy seemed validated when French physician Apollinaire Bouchardat noticed the disappearance of sugar in the urine of diabetic patients whose food had to be rationed during the 1870 German siege of Paris. Nutritional manipulations remained a popular therapeutic strategy through the early 20th century. The approaches varied from starvation or overfeeding to diets based on oatmeal, potatoes or rice.

An effective treatment would have to wait for a better understanding about the cause of diabetes mellitus. A major clue surfaced at the close of the 19th century when German physician Joseph von Mering decided that he wanted to learn about the role of the pancreas in digestion. Apparently unaware of the many failed attempts to remove a pancreas from an experimental animal, von Mering's colleague Oskar Minkowski performed such an operation on a dog. A few days later, Minkowski noticed that the dog – although housebroken – repeatedly urinated on the laboratory floor. Knowing that frequent urination is a symptom of diabetes, Minkowski tested the dog's urine and found it high in sugar. Minkowski and von Mering suspected that the removal of the pancreas had created diabetes. If so, then the pancreas must produce an anti-diabetic substance. The trick would be to find it.

Insomnia Awakens Insulin Therapy

On July 1, 1920, Dr. Frederick Grant Banting began his medical practice in London, Ontario. It did not go well. By the end of the first month, he had seen one patient and had earned four dollars. Business improved little over the next several months. To supplement his income, Banting obtained an appointment with the University of Western Ontario as a demonstrator in surgery and anatomy.

During the night of October 30, Banting prepared a lecture about the relationship between diabetes and the pancreas. He then read a review article by the University of Minnesota's Dr. Moses Barron, who described patients with stones in their pancreatic ducts. The blocked ducts caused the pancreas to atrophy and yet, the patients had not developed diabetes. Autopsies revealed that the pancreases had not totally degenerated; specialized tissue called the islets of Langerhans had remained intact. Unable to sleep, Banting wondered about an experiment. Could he induce a similar type of atrophy by tying a dog's pancreatic duct? Could he then isolate an anti-diabetic substance from the degenerated pancreas? Early in the morning, he wrote a note to himself about the experiment that changed his life and the lives of diabetics.

Banting discussed his idea with Professor John James Rickard Macleod, the head of the physiology department at the University of Toronto. Macleod greeted Banting's proposal with skepticism. What did Banting hope to accomplish when the best-trained physiologists had failed to prove that the pancreas excreted an anti-diabetic substance? But in the face of Banting's persistence, Macleod relented. "Negative results would be of great physiological value," he assured Banting.

Macleod provided a small laboratory and two graduate students, who would take turns helping with the experiments. When work began in May 1921, Charles Herbert Best and Clark Noble flipped a coin to see who would assist Banting first. Best won the toss and became Banting's first and only assistant.

To create a source of anti-diabetic substance, Banting and Best tied the pancreatic ducts of dogs. They removed pancreases of other dogs to create diabetic animals. Late in July, they prepared a pancreatic extract and injected it into a diabetic dog. The injected extract decreased the level of blood sugar, caused a disappearance of sugar from the urine and increased the dog's survival time. Banting and Best worked over the next several months to improve the extract, which they called "Isletin."

Although Macleod saw promise in the results, he insisted that the studies had to be expanded to generate additional confirmatory data. Banting needed a greater supply of Isletin and decided to use the pancreases of unborn calves obtained from beef cattle at William Davies Company's abattoir in Toronto. Later, Banting and Best used the pancreases of mature cows and pigs to prepare their extracts.

By November, the researchers reported that diabetic dogs treated with the pancreatic extracts had markedly reduced blood sugar and sugar secretion in the urine. Encouraged by the studies, Macleod obtained the help of Dr. James Bertram Collip, a University of Alberta biochemist, who isolated a purer product. Soon, they were ready to test a beef extract on a human.

The first person to receive Isletin was 14-year-old Leonard Thompson, a subject of starvation treatment who weighed 65 pounds on admission to the hospital. On January 11, 1922, Thompson received an injection of the extract. It did not produce promising results: a slight decrease in blood sugar and glucose excretion, and an abscess caused by impurities. Collip refined his purification process and within two weeks, Thompson received injections of improved extract. Not only did Thompson's clinical symptoms improve, but, as Banting later reported in his 1922 *Canadian Medical Association Journal* article, the "boy became brighter, more active, looked better and said he felt stronger."

Success brought a new challenge to the Toronto group: to manufacture large amounts of the anti-diabetic substance, now named insulin. In the initial plan, the University of Toronto's Connaught Anti-Toxin Laboratories would finance and supervise insulin manufacture. But technical difficulties in scaling up production and the pressure to mass produce large quantities of insulin shortly led to a deal with the US firm, Eli Lilly Company (Indianapolis, Indiana). Eli Lilly would provide additional financial backing, staff and pancreatic tissue from the Chicago stockyards.

By the end of 1923, insulin was commercially available in Great Britain, Canada and the United States. One year later, insulin became available worldwide.

A Legacy

"Insulin: Its Real Value," an article published in the August 4, 1923 edition of *The Illustrated London News*, concluded with these thoughts:

Whether, by thus alleviating the symptoms of the sufferer, this treatment will allow the pancreas to rest, rehabilitate itself, and lead to cure; or whether the cause of the disease, which is still unknown, will progress further despite the treatment, and create troubles at present unthought of – are problems which time and experience alone can reveal.

Clinicians had hoped that insulin might cure diabetes, but they soon learned that the treatment did not transform an impaired metabolism. Insulin therapy did offer a different type of transformation: It brought the chance for long term survival to those who had a previously fatal disease.

Today, a diabetic can self-medicate with a pre-filled, dose-adjustable pen that injects pure, genetically-engineered human insulin. No longer at the mercy of starvation diets and other experimental remedies, diabetics can manage their own treatment. This independence is another legacy of Banting's achievement.

Miasma Theory

In his treatise, *On Airs, Waters and Places* (400 BC), Hippocrates suggested a connection between disease and poorly drained land with its dank and foul air. A theory developed that disease could be spread by inhaling vapors, or miasma, emitted by rotting animal and vegetable matter. In the 5th century BC, Plato described miasma as vapors released from putrefying bodies. During the 2nd century, Galen warned about the danger of inhaling the foul-smelling breath of ill people.

Sir John Pringle echoed Galen's warning in the 18th century. He said that the bad air and effluvia in hospitals numbered among the main causes of sickness. This hazard, he urged, could be eliminated by cleanliness, good ventilation and the wide spacing of beds. A century later, Florence Nightingale, who also believed in a connection between disease, filth and foul odors, campaigned for clean hospitals.

In the 19th century, the view of the miasmatists seemed confirmed by disease-ridden European cities where crowding coupled with a lack of sanitation. "No fever produced by contamination of the air," said English physician Thomas Southwood Smith, "can be communicated to others in a pure air."

The miasma, or pythogenic, theory encouraged the public health movement and sanitary reforms. Civil engineers listened to miasmatists when they planned new drains and sewers. In North America, cremationists urged a reform of burial practices.

Toward the end of the 19th century, a growing number of contagionists challenged the miasma theory. They believed in Louis Pasteur's germ theory: Small organisms spread infectious disease, not foul air. John Snow, a leading proponent of the germ theory, showed that cholera had spread through central London in water.

The contagionists triumphed over the miasmatists. Yet the miasmatists earned a victory of their own. Fearful of foul air, miasmatists created a sanitation revolution, even if it was for the wrong reason.

Caisson Disease

The demand for coal skyrocketed during Europe's industrial revolution, and inventors sought ways to extract coal from previously inaccessible sites. In 1839, Charles-Jean Triger, a French mining engineer, devised a method of excavating a rich vein of coal buried under quicksand at Chalonnes. Triger had a 70-foot long metal tube – a caisson – built in a mine shaft. Using compressed air, he drove water and sand from the caisson and cleared a space for miners. The miners entered and exited the pressurized shaft through an airlock on the surface.

A French company used a caisson to evacuate underground water from a mine. The technology overcame the difficulty of installing a drainage machine with a capacity greater than that of incoming water currents. The new technique had a price: Miners experienced severe pain in their arms, chest and legs. Two died.

In 1854, Pol and Watelle published a treatise on the effects of air compression. The danger does not lie in entering a mine under compressed air, they observed. "You only pay when you leave."

British engineers adapted Triger's caisson technology to sink piers that supported bridges. Workers build the shell of a pier in a large box or tube, which was closed at the top. After sinking the caisson – open end first – into the river bed, compressed air forced water from the caisson. Toiling in pressurized air, the men dug to the bedrock. To form the pier's core, workers poured concrete into the caisson's interior.

While traveling through Europe, James Buchanan Eads observed the use of compressed air caissons. In 1869, he used the technology to build the first bridge across the Mississippi River at St. Louis. Problems arose. At about 55 feet below the river's surface, men had joint pains while they dug out the riverbed. As they continued to dig, they experienced painful paralysis of their legs. Eleven caisson workers died in five months. Eads shortened the worker's hours and called for his personal physician, Alphonse Jaminet. Despite the doctor's efforts, another nine men died from compressed air exposure.

In September 1871, Washington Roebling started work on the New York caisson of the Brooklyn Bridge. Borings indicated that bedrock lay from 77 to 92 feet below the surface. Every two feet translated into an additional pound of pressure

Mouth of the supply shaft of the Brooklyn caisson of the East River Bridge from *Frank Leslie's Illustrated Newspaper* 31:76 (October 15, 1870). Source: Library of Congress, Prints & Photographs Division, [LC-USZ62-124944].

By late January, the caisson reached a depth of 51 feet, and workers experienced serious health problems. Roebling commissioned Dr. Andrew H. Smith to help his men. During the next four months, Smith, who coined the term "caisson disease," treated men for the mysterious illness.

On May 18, 1872, the caisson reached a depth of 78 and one-half feet. Three men had died from caisson disease. Aware of the St. Louis Bridge tragedies, Roebling stopped excavation. He gambled that the tower on the New York side could stand on sand.

Smith treated 110 reported cases of caisson disease. Undoubtedly, the disease had affected many more men, who feared that a reported illness would ban them from future projects. Workers kept to themselves if they had a dose of the Grecian Bends, a name shortened to the "bends" and inspired by the Grecian Bend, a fashionable forward bent posture of women who wore corsets, crinolettes and bustles.

Smith, Pol and Jaminet correctly focused on rapid decompression as the cause of caisson disease. After a shift in a caisson, the drop in atmospheric pressure caused dissolved nitrogen to form bubbles in the workers' bodies. A rapid decompression overwhelmed their lungs' ability to expel the gas. Consequently, nitrogen bubbles hindered the flow of oxygen to tissues, causing intense pain and even permanent damage.

Treatment for Rickets

Humans can acquire sufficient vitamin D by eating certain foods or by synthesizing it. Ultraviolet light converts a precursor compound in human skin to vitamin D3, which the liver and kidneys metabolize to vitamin D's active form, a hormone that regulates the uptake of calcium and phosphorous. Vitamin D deficiency causes several disorders, including rickets, a disease characterized by soft bones, muscle spasms, seizures and infant death.

Rickets plagued the population of 17th century England. The disorder afflicted so many, that it became known as the English disease. The word "rickets" first appeared in print in the 1634 *Annual Bill of Mortality* for the City of London. Between 1645 and 1668, Daniel Whistler, Arnold Boote, Francis Glisson and John Mayow published unambiguous descriptions of the disease.

Reduced exposure to sunlight can explain the prevalence of rickets among those who lived within the crowded, walled City of London with its narrow streets and oppressive air pollution. A poor diet can also lead to rickets, which may explain the appearance of the disease in rural southern England and the West Country.

One of the treatments for rickets reflected the cure-all of the time: Bleed the patient to draw out bad humors. A vein in the ear lobe proved a popular target. In his treatise, Glisson also suggested cauterization and blistering. Consumption of crow livers offered a less violent treatment for rickets, at least from the perspective of humans.

Cod-liver oil treatment gained momentum during the 19th century. In 1824, the German medical literature contained reports of the oil as a remedy for rickets. Armand Trousseau of France also urged a cod-liver oil regimen, and in 1861, suggested that rickets arose from a lack of sunlight exposure and a flawed diet.

North America was no stranger to the disease. Rickets went unchecked in the New England colonies of the 17th and 18th centuries. Early treatments varied. In one strategy, a mother fed her baby with a paste made from snails, earthworms, ale, herbs and spices. Another remedy called for a mother to dip her infant in cold water, warm the infant in a cradle and then draw blood from its feet.

Eventually, cod-liver oil therapy arrived in the United States. "[I]t is in rickety patients," *Harper's New Monthly* (May 1873) informed readers, "that cod-liver oil has its most positive and curative action."

Based upon correspondence with medical missionaries across the globe, England's Theobald Palm concluded that a deficiency of sunlight caused rickets. In 1890, he recommended "the systematic use of sun-baths as a preventive and therapeutic measure in rickets." Clinical studies of the early 20th century showed that infantile rickets could be cured with exposure to sunlight or ultraviolet rays from carbon-arc and mercury-quartz lamps.

Harriette Chick and her colleagues in Vienna investigated the preventive and therapeutic uses of cod-liver oil and sunlight for infantile rickets. During the early 1920s, they confirmed the value of both treatments.

Researchers explored the connection between these disparate cures by irradiating various foods and excised skin samples. They discovered that irradiation produced a substance that treated rickets as effectively as the cod-liver oil factor, which had been named vitamin D.

The development of vitamin D production ushered the introduction of vitamin D-fortified margarine in 1927. Four years later, the first vitamin D milk appeared on the market; the status of unsavory cod-liver oil suffered. Milk-fortification continues to provide a significant source of vitamin D in Canada and the United States.

The Iron Lung

Scholars point to Biblical passages as evidence that the technique of mouth-to-mouth resuscitation has been practiced for thousands of years. For a comparatively brief time, mechanical devices have been used to force air into someone who has ceased breathing.

Paracelsus, a 16th century German-Swiss alchemist and physician, found a handy tool for positive air pressure ventilation: a fireside bellows. He would insert the tip into the nostril of a patient who had trouble breathing and pump air into the lungs. Apparently, Paracelsus employed secondhand bellows contaminated with cinders. This might explain why his technique had little success.

During the 17th century, the Dutch refined the bellows method to revive drowning victims pulled from canals. For the next two hundred years, European physicians and researchers tried various methods for performing artificial respiration with bellows, pistons and other positive pressure devices. Typically, the methods required an operator to manually sustain a rhythmic inflation of the patient's lungs.

In the early 19th century, medical practitioners raised doubts about the safety of positive pressure ventilation. By the mid-1800s, researchers found an alternative method for drawing air into the lungs: a negative pressure ventilator that alternates subatmospheric and atmospheric pressures around the chest and abdomen. Under negative pressure, the lungs expand and draw in air. With a return to normal air pressure, the passive recoil of the chest allows expiration of air.

John Dalziel of Scotland reported the use of a negative pressure tank respirator to ventilate an apparent drowning victim in 1838. Fifty years later, New York's Joseph Ketchum devised his pneumatic cabinet, an airtight box in which a patient could sit and breathe air from outside through a flexible tube connected to a mask. An operator pulled out or pushed in a large rubber membrane that covered an opening in the box to create low density or compressed air within.

South African W. Steuart developed a sealed wooden box with a motor-driven bellows to treat children with poliomyelitis during the early 1900s. The box enclosed all but the head of the child. An electrically driven pump produced positive and negative pressures inside the box.

In 1926, the Consolidated Gas Company of New York established a committee to fund research for treating respiratory failure caused by carbon monoxide poisoning, drowning and electrical shock. The committee decided to support experiments on artificial resuscitation performed by Philip Drinker and Louis Agassiz Shaw at the Harvard School of Public Health. Although Drinker and Shaw built their machine to sustain breathing for those injured by industrial accidents, it played a critical role in the treatment of poliomyelitis victims.

The Drinker-Shaw apparatus enclosed the patient's body in a cylindrical sheet-metal tank sealed at one end. At the other end of the tank, the patient's head and part of the neck protruded through a rubber collar that created a seal for an airtight chamber. While the patient's head remained at atmospheric pressure, a pump raised and lowered the air pressure around the body. Low pressure allowed the chest to expand and air to rush into the lungs. Positive pressure compressed the chest and forced air from the lungs.

"By this method," Drinker and Shaw wrote in their June 1929 report, "movement of the chest is induced in such a manner as to simulate the natural respiratory movements."

Drinker nicknamed their invention the "iron lung."

John Haven Emerson ran a Harvard Square machine shop that produced medical and scientific instruments. After he saw the Drinker-Shaw iron lung, he began to devise improvements. Three years later, he had developed a simpler, quieter and less expensive apparatus.

Emerson's efforts brought a lawsuit alleging infringement of Drinker's patents. Through an extensive search of engineering and medical literature, Emerson and his colleagues showed that the components of the iron lung had been known before Drinker and Shaw filed their patent applications.

A Massachusetts federal district court judge declared the Drinker patents invalid. Emerson's iron lung prevailed in the marketplace.

Iron lungs saved the lives of many victims of polio epidemics from 1930 to 1960. During the last major North American epidemic of the 1950s, hospital polio wards carried 50 to 100 operating iron lungs.

By the early 1960s, the polio vaccines of Jonas Salk and Albert Sabin radically diminished the occurrence of the disease. Improved positive pressure ventilation devices replaced the negative pressure ventilation technique for the treatment of other respiratory disorders. In 2004, about 40 US polio survivors still depended on their iron lung machines.

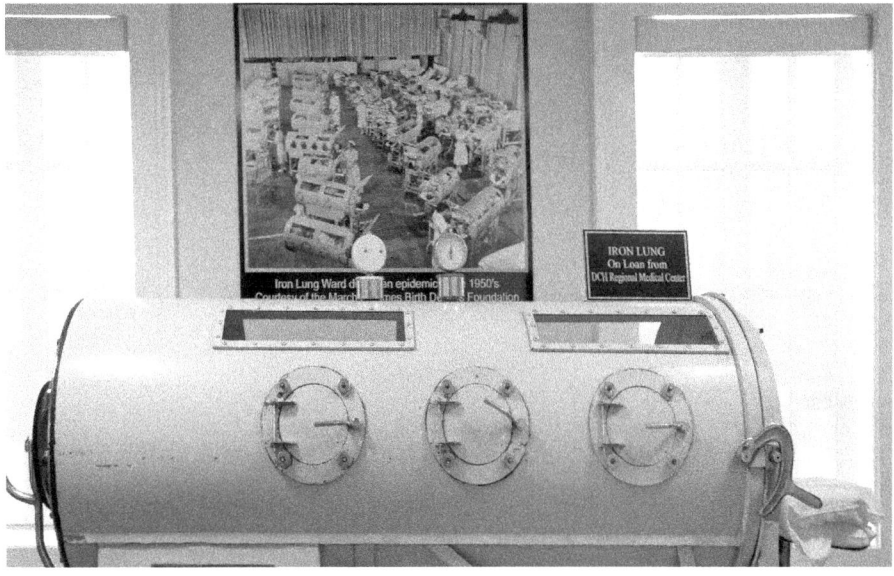

An iron lung, manufactured around 1933. The poster behind the iron lung shows an "iron lung ward" in a hospital during a 1950s epidemic. Source: The George F. Landegger Collection of Alabama Photographs in Carol M. Highsmith's America, Library of Congress, Prints and Photographs Division.

Revigator

"I have improved from the first day I began using the water and gained six pounds in weight," said J.D. Bright in his testimonial of March 30, 1921. "My digestion is now perfect and I am sleeping sound. I am feeling much better than I have for a long time." The aptly named Mr. Bright extolled the virtues of radium water. That's right – radioactive water.

During the early 20th century, radium contamination imbued spring waters with a new glow. In 1914, for instance, promoters at Hot Springs, Arkansas, capitalized on the discovery of radium in their spring water by boasting about the radioactivity of the waters. They informed the public that radioactive substances carry electrical energy deep into the body, stimulating secretory and excretory organs at the cellular level and hastening the exit of waste from the body.

Not everyone could invest the required time and money for a pilgrimage to a radium hot spring. Why not produce radioactive water at home? The 1920s experienced an explosion of radium water dispensers. The Radium Ore Revigator, the Vitalizer, the Radiumator, and other containers enjoyed millions of buyers across the country. Advertisements bombarded the public with the notion that drinking "liquid sunshine" exposed internal organs to the sun's healing rays.

Perhaps the most famous water dispenser was produced by The Revigator Water Jar Company of San Francisco, which sold several hundred thousand Revigators. To promote their product, the company published a red-covered booklet entitled *The Revigator Water Jar For Every Home*. Here, the company explained that a "surprising amount of illness is caused from drinking improper water." It seems that natural water discharged from the earth is radioactive, whereas ordinary drinking water does not possess the property to any appreciable extent. Lined with a uranium ore, the Revigator released radon gas that dissolved in the stored water. In this way, the "ore continuously revigorates or restores natural vigor to drinking water placed therein." As a bonus, "Revigator water tastes softer, more palatable and less heavy."

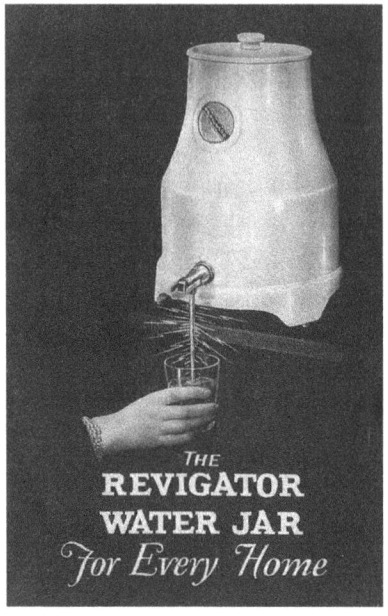

Cover of a Revigator Water Jar Company booklet.

Health-conscious consumers heard endorsements for trendy radioactivity treatments from the media, physicians, and scientists. Acclaim was not universal, as shown by the commentary "Radium Ore Revigator" published in the November 21, 1925 issue of *The Journal of the American Medical Association*. "A huge traffic has been developed during the past four or five years in the sale of so-called radioactive pads," the authors noted, "consisting of a few cents' worth of crude ore having a low grade of radioactivity and possessing no more therapeutic value than do the luminous figures on the dial of a two-dollar watch." Yet a few dissenting voices failed to tarnish the aura of the faddish remedies.

In 1932, industrialist millionaire and socialite Eben M. Byers died after suffering a prolonged and mysterious illness that caused his bone marrow and kidneys to fail, and his bones to splinter and break. By the time that he died, the former sportsman weighed ninety-two pounds; he had anemia, a brain abscess and advanced decomposition of the jaw. For four years, Byers had guzzled as many as 1,500 bottles of Radithor, an expensive blend of radium-226 and radium-228 in distilled water. The medical community and the press pointed to Radithor as the cause of death. Although the public shunned the brand of liquid sunshine, the popularity of radium water did not suffer. However, the dream of a radioactive enhancement of health decayed after the public became aware of the effect of an atomic bomb.

Penicillin

Penicillium, a blue-green mold, secretes compounds lethal for certain bacteria. Although the practice of using molds as therapeutic agents dates at least to ancient Greece and India, the addition of *Penicillium* toxins to the healer's arsenal required several rediscoveries and a world war.

Interest in the antibacterial properties of certain molds gained momentum during the 19th century. In 1871, English surgeon Joseph Lister discovered that a mold inhibited bacterial growth in a urine sample. Four years later, John Tyndall, an Irish scientist, reported that a *Penicillium* species caused bacteria to burst.

In 1897, French medical student Ernest Duchesne completed his doctoral dissertation on competition among microorganisms. Duchesne described how *Penicillium glaucum* destroyed bacteria in a culture, and he reported that the mold's secretions protected experimental animals from a lethal dose of typhoid bacteria. "One might thus hope that by pursuing the study of biological competition between molds and microbes," Duchesne wrote, "one might be led to the discovery of other facts which would be directly useful and applicable to prophylactic hygiene and to therapy." For three decades, Duschesne's hope lay as dormant as a mold spore.

In 1928 Alexander Fleming, a Scottish bacteriologist at St. Mary's Hospital in London, rediscovered *Penicillium*'s antibacterial properties. According to one version of the discovery, Fleming returned from a vacation to his typically chaotic laboratory around early September. While clearing a pile of used Petri dishes, he studied one dish and said, "That's funny."

What Fleming found odd was the appearance of bacteria colonies spread across the plate except near a bit of mold, which he identified as a *Penicillium*. Fleming grew the mold in a pure culture medium, and confirmed that the mold inhibited the growth of particular bacteria. Further experiments showed that the mold secreted a chemical, dubbed "penicillin" by Fleming, which not only inhibited bacterial growth, but also killed bacteria. Yet even large doses of penicillin did not produce toxicity in experimental animals.

Fleming published his results in 1929 and presented his findings at a seminar. Despite his peers' resounding lack of interest, Fleming continued his studies and tried to get chemists to purify penicillin. Soon after one chemist characterized the production of penicillin for therapeutic use as almost impossible, Fleming abandoned efforts to purify the antibiotic.

During the summer of 1938 at Oxford University, German biochemist Ernst B. Chain discovered Fleming's paper while reviewing scientific literature. Chain showed the paper to a colleague, Australian-born Howard Florey, and the two scientists decided to give penicillin a closer look. In May 1939, while British troops prepared to evacuate Dunkirk, Florey's group injected mice with a virulent strain of streptococcus. Then, they treated half of the mice with penicillin. The penicillin-treated mice survived.

To boost antibiotic production for clinical studies, Florey turned to the British pharmaceutical industry. In the midst of war, however, companies lacked personnel, material, or funds for Florey's project. Deciding to try the United States, Florey and coworker Norman Heatley concealed vials of freeze-dried mold in their baggage and flew to Lisbon in a blacked-out airplane. From Lisbon, they traveled to New York and arrived on July 3, 1941. Meetings with representatives of private and government organizations produced a recommendation to try the US Department of Agriculture laboratory in Peoria, Illinois. The USDA lab had an enormous fermentation facility built for investigations of new uses for agricultural products.

In Peoria, Heatley and Andrew Moyer discovered that large quantities of *Penicillium* could be grown in a liquid culture medium. Unfortunately, the *Penicillium* strain produced mediocre quantities of penicillin. A worldwide search for a superior strain came up empty. Then, a lab assistant, nicknamed Moldy Mary, brought a rotten cantaloupe that she had found at a Peoria market. From the fruit's moldy stem, scientists isolated a strain of *Penicillium* that became the origin of most of the penicillin produced in the world.

With the Second World War raging, penicillin found immediate use in the military. In 1943, Florey and a coworker used penicillin to treat British army soldiers in North Africa, while American doctors treated soldiers in the Pacific theater. By the invasion of Normandy in June 1944, pharmaceutical companies produced 100 billion units of penicillin per month.

This poster urged efficiency in construction of a penicillin manufacture plant, circa 1941-1945. Source: US National Archives.

During the late 1950s, scientists achieved an objective that had eluded a secret, massive international war effort: the chemical synthesis of penicillin. Since then, chemists have invented penicillin derivatives with improved properties. The ongoing search for new antibiotics is driven by the development of antibiotic-resistant bacteria, a phenomenon that Fleming predicted in his 1945 Nobel Prize lecture.

Security Coffins

The plagues of 17th and 18th century Europe required hasty burials that brought the risk of burying live plague victims. Even without the panic of an epidemic, premature burial remained a possibility in a time that predated routine embalming and reliable indicators of death. Numerous anecdotes about accidental burials fed fears about this horrible fate.

During the late 18th century, many German states offered one way to prevent premature burial: the waiting mortuary, or *Leichenhaus*. Here, the presumed dead would be held until the onset of putrefaction signaled that an individual had truly departed and did not linger in *Scheintod*, the death trance. However, momentum for the *Leichenhaus* movement soon died. People refused to bring their deceased loved ones to a foul-smelling "Asylum for Doubtful Life."

The Industrial Revolution ushered in an era of dangerous machinery, increased chemical use, and experiments with electricity, factors that created new ways to induce death-like comas in accident victims. The Revolution also brought an inventiveness that spawned alternatives to the objectionable *Leichenhaus*: security coffins. In 1792, Duke Ferdinand of Brunswick designed a coffin for interment in a vault. He equipped his coffin with a window, an air hole, and a locking lid. A special pocket in the shroud contained a pair of keys, one for the coffin and one for the vault door.

Since few could afford a burial chamber, P.G. Pessler, a German village parson, proposed attaching a hollow tube to buried coffins. A rope to a nearby church bell would pass through the tube to the inadvertently interred who would have to muster sufficient strength to ring the bell. In the second half of the 19th century, German inventors patented about 30 versions of security coffins. One included a pyrotechnic rocket that would be launched through the coffin's tube. Another design had a coffin tube with a loud siren mounted inside.

In Great Britain, those who wished to avoid premature burial could purchase coffins outfitted with a Bateson Life Revival Device. An iron bell in a miniature tower, or Bateson's Belfry, would be mounted on a coffin's lid, and a rope attached to the bell would run through a hole in the lid to the supposed deceased's hands.

Franz Vester filed the first US patent application for a security coffin in 1868. His coffin lid had a sliding door fitted with a detachable, hollow cylinder. Upon recovery, the prematurely buried individual would crawl through the coffin opening, climb a ladder bolted inside and open a door to the surface. Other patented coffins had electrical signals that triggered flags, bells and rotating lights. Some conveniently offered an electric light, a heater and a telephone.

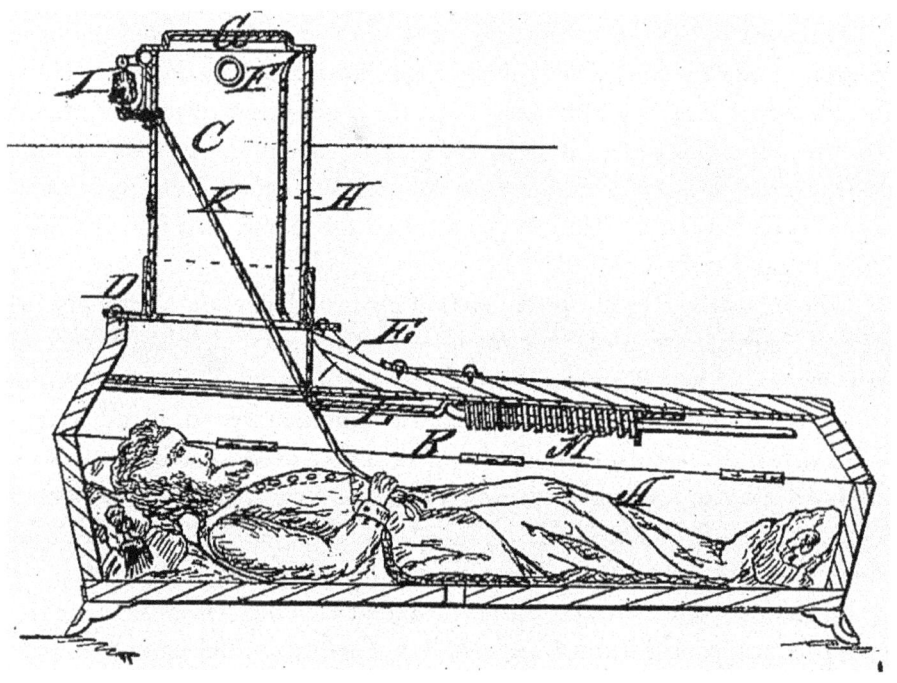

Illustration from US Patent No. 81,437 (1868), Franz Vester's "Improved Burial-Case."

Fear inspired the invention of security coffins and fear might have prevented their mass production. An editorial in the January 29, 1898, issue of the *Journal of the American Medical Society* highlights that a coffin's limited air supply should prevent any return to consciousness. "It is no comfort to think that any one can be buried alive," the editor observed, "but it is better to believe, as the facts warrant us, that the long lingering death in hopeless horror, and the powerless anticipation of the fate are still less probable events." Yet the majority of late 19th century security coffins provided an unlimited air supply. Jan Bondeson, author of *Buried Alive* (2001), suggests that the intended customers of security coffins, fearful of premature burial, also feared that the coffin's alarm mechanisms might not function properly. With an adequate air supply and a defective signaling device, the prematurely buried could fully experience the horrible fate that they had attempted to avoid.

Safety Razor

As early as 30,000 BC, men faced the dicey task of shaving. In Stone Age cultures, men scraped facial hair with flints, shells, shark's teeth, or obsidian glass. Razors made of copper, bronze or gold became available after the invention of metal smelting. The Romans devised the appropriately-named cut-throat razor with a long, one-sided straight-edged blade.

In 1762, French barber Jean-Jacques Perret invented a simple safety razor with a blade sheltered within an L-shaped wooden guard. The design did not prevent nicks, but it did stop the blade from slicing too deeply. During the century, others incorporated a skin guard in a razor. But the user still had to remove the blade for sharpening.

One day in 1895, traveling salesman William Gillette talked with William Painter. Why not invent something perfect for a salesman, Painter suggested, something disposable that had to be purchased repeatedly, something with built-in obsolescence. Painter knew about such inventions; he had invented the disposable crimped bottle cap.

The idea of a disposable blade came to Gillette while shaving. He built a crude device that housed a blade in a way that made it difficult for a shaver to cut himself. Available technology did not enable Gillette to make inexpensive, thin, sharp blades. He experimented for years.

By 1902, Gillette devised a process for making and sharpening steel blades, and T-shaped handles to house the blades. Gillette's company produced 51 razors and 168 blades in 1903. Five years later, the company sold 300,000 razors and 14 million blades. The company supplied the US Army during the First World War and by the end of the war, people around the world knew about the Gillette razor.

The Ocular Prosthesis

The crafting of artificial eyes has fascinated humans for thousands of years. In the 5th century BC, Egyptian priests made ekblaphara, a synthetic eye of enameled metal or painted clay attached to cloth and worn in front of the eyelid.

By the 15th century, artisans made hypoblaphara: artificial eyes designed to rest under the eyelid. In 1561 French surgeon Ambroise Paré described a gold and colored enamel artificial eye for placement beneath the eyelid in an empty socket or on top of an eye's atrophied remains.

Sixteenth century Venetian artisans formulated a type of glass that could be tolerated in the eye socket. Although considered to be among the finest in the world, Venetian glass eyes tended to be very fragile and uncomfortable.

During the early 19th century, enamel dominated ocular prosthetic manufacture, despite high cost and lack of durability. Ocularists abandoned enamel soon after 1835 in favor of cyrolite glass introduced by German artisans. Craftsmen painted cryolite glass eyes to match the natural colors of the patient's eye.

By the end of the 1800s, immigrant German ocularists brought the art of glass eye making to North America. Yet cryolite glass continued to be exported from Germany.

The commencement of the Second World War ended supplies of German cryolite glass. To aid wounded soldiers, the US government searched for a replacement. In 1944, the US Army's Dental Corps reported the successful application of dentistry plastics technology to fabricate molded eyes composed of medical-grade acrylic.

Although cryolite glass eyes are still used in many parts of the world, the vast majority of patients wear acrylic ocular prostheses. Acrylic can resist broad temperature fluctuations, it is lightweight yet practically unbreakable, and acrylic can be molded for increased comfort.

Morphine – The Divine Drug

Lance the pods of the poppy, *Papaver somniferum*. A liquid oozes from the wound. Collect the liquid and dry it. The resultant brown resin is opium.

Humans have applied this basic process for millennia. Around 4000 BC, Sumerian cultures used opium as a narcotic. They called the poppy Hul Gil, the "joy plant." By the first and second centuries AD, Græco-Roman works included descriptions of opium as a medicine.

During 17th century Europe, laudanum became one of the most popular opium-based medicines. According to the *London Pharmacopæia* (1618), laudanum was a pill made from opium, saffron, castor, ambergris, musk and nutmeg. These "stones of immortality" relieved pain, coughing and diarrhea. Laudanum remained popular through the 1800s, a time when people faced bacteria-infected drinking water, skies thick with coal smoke, and regular assaults of cholera and dysentery.

Concocting a medicine with opium required guesswork; the potency of opium could vary six-fold from batch to batch. Chemistry brought consistency.

In the early 1800s, German apothecary Friedrich Wilhelm Sertürner isolated a compound from opium that induced drowsiness in dogs. Sertürner and three youths tried the material and experienced the symptoms of severe opium poisoning. The chemist had discovered opium's active principle. Sertürner named it morphium after Morpheus, the Greek god of dreams. Later, others renamed the compound, morphine.

French physician François Magendie stimulated interest in the medicinal use of morphine. In 1818, he reported that morphine had relieved a patient's pain and allowed her to sleep. During the 1820s in Europe and the 1830s in North America, commercial manufacture of morphine made the drug available in standardized doses.

Morphine was not the easiest substance to administer. A morphine paste could be introduced into the body with the point of a lancet. Alternatively, a solution of morphine could be dripped into a wound.

In 1853, Charles-Gabriel Pravaz in France invented the hypodermic syringe, a device refined two years later by Alexander Wood in England. The syringe greatly facilitated morphine administration. Morphine injections were used for localized pain, and to prepare a patient for surgery under general anesthesia.

During the American Civil War, physicians relied on injected morphine as a surgical anesthetic. Wounded soldiers returned home with a supply to relieve their pain. By the end of the war, hundreds of thousands had morphine addiction, the "Soldier's Disease." Europe's Franco-Prussian War also inadvertently created morphine addicts.

At home, doctors prescribed morphine to treat pain, rheumatism, headaches, opium addiction and a variety of other conditions. Many soothing elixirs boasted morphine as the active ingredient. Mrs. Winslow's, the most popular American teething potion, contained sugar syrup, fennel, anise, caraway, alcohol and morphine sulfate. Other teething remedies also delivered children into the arms of Morpheus.

Although dosing an aggravating child with a narcotic may be frowned upon today, morphine was considered to be safe and reliable. It was hailed as "God's own medicine."

Tripping on Winslow's syrup. Source: An 1886 advertisement-calendar card.

The attitude about morphine use altered toward the end of the 1800s. Drug addiction had become a recognized problem. Now, chemists searched for alternatives to morphine.

In 1874, English chemist C.R. Alder Wright synthesized diacetylmorphine in his search for a non-addictive substitute for morphine. Twenty years later, Heinrich Dreser, who worked for The Bayer Company of Elberfeld, Germany, devised a commercial process for producing diacetylmorphine, a powerful painkiller that seemed to lack morphine's side effects. Bayer marketed diacetylmorphine as an analgesic and cough suppressant. They named it "Heroin," apparently based on the German word for heroic. As it turned out, heroin failed as a safe morphine replacement. The company eventually dropped the product in favor of aspirin.

Did heroin offer an effective step-down cure for morphine addiction? Physicians debated this topic in the early 20th century. Some noted that their patients suffered withdrawal symptoms with heroin equal to those experienced with morphine withdrawal. It seemed that heroin did not offer a bona fide cure for morphine addiction any more than morphine cured opium addiction.

During the first decade of the 1900s, Canada and the United States enacted laws that effectively limited narcotic content of products sold as medicine. In 1914, the US Harrison Tax Act served to make illegal any nonmedical use of morphine. Within a decade, the US Treasury Department enforced limitations on the import and sales of narcotics.

Today, physicians use more than 230 tons of morphine every year to relieve chronic pain and post-operative pain, as well as other medical purposes. Chemists have invented many synthetic pain relievers. Yet morphine remains the standard against which new analgesics are compared.

IV. Communication

Pencils

A violent storm blew through Borrowdale, England one night around 1500. According to legend, when shepherds searched for their sheep the next morning, they found that the wind had torn trees from the ground, exposing a black material that had lain hidden under the soil. The substance looked like coal, but would not burn like coal. The strange material became known as plumbago or black lead. Centuries later, it would be renamed graphite.

Shepherds found the material useful for marking their sheep. Others used graphite for treating colic, easing the pain of urinary disorders and manufacturing cannon balls. Flemish traders have been credited with spreading the idea that graphite could be a useful material for artists. Graphite-based pencils eventually gained popularity and replaced lead-alloy styluses. Among its advantages, graphite left a mark darker than lead, a mark that could be erased by rubbing with bread crumbs.

Early graphite writing tools consisted of rope or string wrapped around a stick of graphite. Wood-cased pencils were made by shoving a graphite stick into the end of a hollow twig or into a piece of wood that had been hollowed out by hand. Later, sticks of graphite were glued into cedar cases. The firmly encased graphite could be whittled to a fine point with a knife. There was little motivation to change the method of producing pencils so long as pure graphite remained available.

In 1793, war broke out between Britain and France, severing the supply of high-quality British graphite. The French Minister of War commissioned Nicolas-Jacques Conté to develop an alternative to using Borrowdale graphite for pencils. Conté, an engineer, inventor and artist, devised a substitute for British graphite by pressing a mixture of wet powdered graphite and clay into square grooves cut into boards. He then recovered the dried leads, fired them at high temperature, and inserted them into wooden cases. Conté's method of kiln firing enabled the manufacture of pencils with varying degrees of hardness, a principle still used today.

A different war kindled innovations in the North American pencil industry. The American Civil War created a demand for pencils that could not be met by a craft-based industry. The need could only be fulfilled by developing automated techniques of mass production, an approach that changed the speed of pencil production and ultimately, pencil shape. Until the mid-1870s, wood-cased pencils carried the traditional rectangular leads. But tradition did not bind America's fledgling pencil manufacturing industry, which adopted new molding machines that produced cylindrical or hexagonal pencils with round leads.

The introduction of round – and centered – leads enabled the development of rotary pencil sharpeners during the end of the 19th century. Sharpened pencils brought a new challenge: The points often snapped inside the wooden shaft and splintered the wood.

In 1933, the Eagle Pencil Company's researchers found that splintering occurred, because the wood and lead lacked a sufficiently strong bond. Manufacturers impregnated pencil leads with wax to impart smoother writing characteristics. The wax also interfered with the glue that bound wood and lead. Eagle Pencil's researchers solved the problem by burning wax from the lead's surface with acid and treating the wood with resinous binder to create a tough sheath around the lead.

On August 22, 1938, a *New York Times* editorial lamented that "writing with one's own hand seems to be disappearing, and the universal typewriter may swallow all." Yet it is the typewriter that is disappearing in the wake of the computer. The pencil endures.

Braille – the Tactile Code

Helen Keller's 1929 *New York Times* article paid tribute to the Braille tactile system of reading and writing. "They who once sat brooding through sad, interminable days of emptiness," she wrote, "now look with rapt gaze upon the universe as they read with their eyes in their fingers." The code named after Louis Braille – the foremost medium of literacy for blind people – has been adapted to nearly every language. The origins of Braille's code can be traced to a man with a vision.

In 1771 Paris, Valentin Haüy attended St. Ovid's Fair, where he watched a group of blind men perform slapstick comedy for money. Decked out in dunce caps and colossal cardboard glasses, the men sat before upside down sheet music and created jarring noises with their musical instruments. Formal education, Haüy decided, would provide the means for the blind to better their lives. A way to achieve this eluded him.

The solution presented itself in 1784 when Haüy pressed a silver coin into the hand of François Lesueur, a young blind beggar. As soon as the boy announced the denomination of the coin, Haüy realized that the blind could read by touch. He supported François, while teaching him to read with wooden letters used to form words.

One day, François told his teacher that he could recognize the letter "o" when he touched the reverse side of an embossed funeral card. Haüy tried an experiment: He used the handle of a quill pen to trace letters on a thick piece of paper. François recognized the characters, and Haüy recognized that raised letters could replace the awkward wooden ones.

Haüy continued to recruit new students. With the help of Société Philanthropique, he founded the world's first school for the blind in 1786.

The Royal Institution for Blind Children offered education to all blind children, regardless of gender or family wealth.

The school also produced embossed books by applying wet paper to raised letter forms. After drying, the paper retained tactile shapes of the letters. The students, however, found the books difficult to use, because they had to trace the shape of each large, elaborate letter. This reflected Haüy's bias: A letter easily interpreted by the eyes need only be slightly altered for interpretation by touch.

Louis Braille eventually invented an alternative. In Coupvray, the French village of his birth, three-year old Louis Braille accidentally blinded himself with a knife or awl, while trying to imitate his father, a master craftsman and harness-maker. By the age of five, infection had robbed the sight from his remaining eye. Louis learned to read by feeling upholstery studs hammered into wood in the shapes of letters.

In 1819, ten-year-old Louis and his father took a coach for the four-hour journey to Paris. Louis became the youngest student at the Institution for Blind Children. Here, he learned the difficult traditional skill of reading Haüy's embossed script.

Captain Charles Barbier visited the school in 1823; it changed Louis' life. While serving in Napoleon's army, Barbier had seen soldiers extinguished, because they had revealed their position in the night by lighting a lamp to read a message. He thought that the military needed a tactile system for sending and reading messages. This would be useful not only for nocturnal communications, but also for maintaining daytime contact with artillery crews, blinded by dense smoke in the heat of battle.

Barbier based his "night writing," or sonography, on a 12-dot cell arranged in two columns of six dots. In this phonetic system, each cell of raised dots represented sounds that combined to form words. After the French Army showed a lack of interest, Barbier thought that the blind might benefit from his code.

Louis Braille became fascinated with Barbier's sonographic system, but he also realized that it had problems. Sonography's 12-dot cell was larger than a fingertip could cover and difficult to write with a stylus. Since the cells represented sounds, the code lacked punctuation marks and numbers. Louis experimented with the code to simplify it.

In 1824, Louis presented a new code in which six dots of domino-like cells provided 63 permutations for letters, numbers, punctuation marks and later, musical notes. The cells contained six dots numbered in a specific order, and each dot or combination of dots represented a letter or other character. For example, dot 1 represented the letter "a," whereas dots 1 and 4 represented the letter "c." With this simpler system, a fingertip could cover an entire cell, allowing a reader to move rapidly from cell to cell.

Although the other students developed a liking for Braille's code, the school did not endorse it. Not only would the cost of introducing a new system be prohibitive during difficult economic times, but sighted faculty members lacked enthusiasm for learning such a strange code. Eventually, the school adopted a British tactile writing system invented by John Alston. Unlike Haüy's approach, Alston's used simplified letter forms without swirls or serifs. To enforce the new system, the school's director, P. Armand Dufau, burned embossed books created with Haüy's process, and books printed in Braille's new code. The school's entire library and product of nearly fifty years' work went up in flames. Yet students continued to use Braille's code privately.

In 1854, France accepted Braille as the country's official communications system for the blind. Despite competing tactile writing systems and opposition to a code alien to the sighted, American and British Braille Committees adopted a universal Braille code for the English-speaking world in 1932. Like any living language, Braille evolves. A form of modern English Braille includes contractions that represent groups of letters or frequently-used words. The Computer Braille Code and the Nemeth Code of Braille Mathematics and Scientific Notation provide special symbols that convey technical and advanced mathematical information.

Helen Keller with her hand on a braille book, circa October 28, 1904. Source: US Library of Congress.

Helen Keller wrote that, "I use Braille as a spider uses its web – to catch thoughts that flit across my mind." Today, a blind individual can use computer screen reader software and a tactile Braille display to capture information and ideas traversing Internet's World Wide Web.

Shorthand – A Brief History

For millennia, recorders have relied upon written codes to meet the challenge of faithfully transcribing the spoken word. Around 200 BC, the Roman poet Quintas Ennius formulated a shorthand scheme requiring 1,100 symbols. In 63 BC, Plutarch credited Cicero's instruction of clerks in abbreviated writing for preserving Cato's Roman Senate speech on the punishment of the Catilinarian conspirators.

When Cicero freed the slave Tiro, he inadvertently advanced the art of condensed writing. Marcus Tullius Tiro invented a new shorthand scheme with abbreviations of commonly used words, phrases and sentences. His system of tachygraphy, or swift writing, included a new symbol: the ampersand.

Tiro's shorthand gained popularity during the rise of Christianity. In 96 AD, Pope Clement I appointed shorthand writers for each of Rome's new seven districts. Cyprian, the Bishop of Carthage, decided to increase the accuracy of the abridged writing by adding several thousand abbreviations to Tiro's system. Students of shorthand probably failed to welcome this modification. According to tradition, when Cassian, Bishop of Brescia, refused to sacrifice to the Roman gods, the Emperor condemned him to death and turned him over to his shorthand students. The Bishop's pupils bound their teacher to a stake and stabbed him with their iron styli. St. Cassian is the patron saint of stenographers and court reporters.

As the Roman Empire declined, so did the popularity of shorthand. In the 6th century, Emperor Justinian forbade the use of abbreviated writing. Seven hundred years later, shorthand was considered secretive, diabolical, and a risky art to practice.

A British clergyman brought shorthand back into the light. Timothe Bright, author of *Characterie – An Arte of Shorte, Swifte, and Secrete Writing by*

Character (1588), invented one of the first shorthand systems in the English language.

Others soon devised their own versions of shorthand for English. Stenographer John Willis published a shorthand alphabet of geometric characters in 1602. Samuel Pepys, using Thomas Shelton's shorthand system, recorded his observations of the Great Plague of 1665 and the Great Fire of London. In 1707, Thomas Gurney developed a system of shorthand that Charles Dickens used when he worked as a parliamentary court reporter. Apparently, Dickens did not enjoy the task of learning Gurney's shorthand. "When I had groped my way, blindly, through these difficulties, and had mastered the alphabet," Dickens' David Copperfield complained, "there then appeared a procession of new horrors, called arbitrary characters."

The early 19th century brought experiments with shorthand codes that used symbols to represent speech sounds. These phonetic systems, as opposed to traditional alphabetic schemes, offered several advantages for recording speech: fewer pen strokes, increased speed and greater accuracy.

In 1837, British educator Isaac Pitman introduced his phonography system, a phonetic approach that used geometrical curves and lines in varying lengths and angles to represent sounds. Pitman's brother Benn introduced the system to North America in 1852. Benn Pitman also served as chief of five stenographic reporters who recorded the trial of the Abraham Lincoln assassination conspirators. By the end of the 19th century, most American shorthand writers used the Pitman system. Pitman's remains the most widely used shorthand scheme in English-speaking areas outside the United States.

Irish stenographer John Robert Gregg invented the currently dominant shorthand system in the United States. Gregg's phonetic shorthand system, published in 1888, represents sounds of words with cursive lines, curves and loops. His slanting characters, similar to common longhand writing, enabled writers to record spoken words at improved speeds.

Like the Pitman system, Gregg's scheme has been adapted to many non-English languages. Shorthand has survived stenographic machines, tape recorders, and so far, voice recognition software.

&

The ampersand symbol evolved from the joining of E and T. *Et* is the Latin word for "and." The two letters can be distinguished in early forms of the ampersand. A piece of papyrus dated to around 45 AD bears an E plus T version of the ampersand written in the style of early Roman capital cursive.

The two letters began to merge in a Roman lowercase cursive, a form of writing that became italics. Hasty writing might have inadvertently created early connections between E and T. In time, writers deliberately joined the middle part of the E with the T. By 775 AD, the letter combination had merged as a single character. Printers of the early 15th century eagerly adopted, and sometimes overused, the aesthetically-pleasing ampersand.

The ampersand symbol became embedded in European writing. By the early 19th century, the character earned a place as the last letter of the English alphabet in school primers. Those learning the alphabet would chant at the end, "and, per se and." The phrase, which simply meant "and the character 'and' by itself," became corrupted to "ampersand."

In modern English language fonts, remnants of the letter E reside in the left side of the ampersand. Only a small fragment remains of the T. & that is that.

Ballpoint Pens

Touted as the "fantastic, atomic era, miraculous pen," the ballpoint pens were guaranteed to write underwater and to require an ink refill only once every two years. But the public's enthusiasm ran dry when the leaky pens turned out to be less than miraculous. The ballpoint pen – an invention 60 years in the making – almost became a short-lived fad.

John J. Loud is credited with obtaining the first patent on a pen that had a sphere at the marking point. His US patent, which issued in 1888, shows that he had considered the ballpoint marker to be an improved fountain pen. Others followed with ballpoint writing devices, and like John Loud's invention, these never entered the marketplace. The availability of ink with a suitable viscosity posed a problem: Ballpoint pens leaked if the ink was too thin and clogged if the ink was too thick.

Ink was on the mind of László Jozsef Bíró when he visited a newspaper printing press in 1938. A Hungarian journalist, artist and hypnotist, László noticed that the printing press ink dried quickly and did not smudge. The ink seemed ideally suited for writing. But the viscous printing ink wouldn't flow from a fountain pen's writing nib. It may be possible, he decided, to replace the nib with a rotatable ball bearing that transferred ink from a reservoir to paper as the writer moved the pen across the surface.

László had little time to think about his new pen. He and his brother György – a chemist – fled Europe to escape Nazi persecution in 1939. The brothers settled in Argentina and began experimenting with ballpoint pen designs. With the help of English businessman Henry G. Martin, László and György founded the Eterpen Company to produce their Biro pens.

The British Government took a keen interest in the ballpoint pens. The Royal Air Force – engaged in fighting the Second World War – needed a replacement for navigators' fountain pens that leaked when fighter planes flew at high altitudes. Henry Martin arranged a small-scale production of the pens and supplied thousands of Biros to the Royal Air Force.

The US War Department obtained several Biros, sent samples to manufacturers, and made an offer: The government would buy an improved version of the pen. In May 1945, Eversharp acquired exclusive US manufacturing and marketing rights from Eterpen and began to pour millions of dollars into a rush project to commercialize ballpoint pens. Around the same time, Chicago businessman Milton Reynolds purchased a few Biros on a visit to Buenos Aires. The pens inspired him. Reynolds returned home, created the Reynolds International Pen Company, and beat Eversharp to the market. The Reynolds' Rocket ballpoint pen debuted in New York City at Gimbel's department store on October 29, 1945. By the end of the week, the store had sold 30,000 pens – at $12.50 each.

For a while, sales of the novel ballpoints surpassed those of fountain pens. But the Reynolds' Rocket and the Eversharp CA ("capillary action") did not live up to the advertising hyperbole. The pens leaked, skipped and deposited gobs of ink. By 1951, fountain pens regained the position of first tier product and Reynolds' company collapsed. Eversharp eventually sold its pen division.

The French Baron Marcel Bich helped to revive the ballpoint pen. During the early 1950's, his BIC pen took over the European market. He tackled skeptical US consumers with a television campaign promising that his improved, inexpensive and disposable pen "Writes First Time, Every Time!" People bought it and still are. Today, the BIC Company's ballpoint pen dominates the world market.

Revealing Cryptic History

Demaratus faced a grim challenge in 480 BC. While living in the Persian city of Susa, Demaratus learned that Xerxes planned to invade Greece. Although he had been expelled from Greece, Demaratus remained loyal to his homeland. How could he slip a warning past Persian guards?

The exile concocted a plan to hide his message. After clearing wax from a pair of wooden writing tablets, Demaratus etched his warning about Xeres' plans into the wood. Then, he covered the message with a coating of wax. The apparently blank tablets passed inspection and reached Sparta in southern Greece. Here, Gorgo, the daughter of a Spartan king, discovered the hidden writing. Prepared for invasion, the Greeks repulsed Xerxes' army.

Steganography, the art and science of hiding messages, has been used for thousands of years. In ancient China, messengers swallowed a wax-covered ball of silk that contained a private communication. In the first century AD, Pliny the Elder described how to use fluid from the thithymallus plant, which dried transparently on paper and turned brown with heating. In the 16th century, Giovanni Battista Porta taught a method of concealing messages by writing on the shell of a hardboiled egg with ink made of alum and vinegar. Although the ink penetrates the porous shell without leaving a trace, the message can be read on the surface of the hardened egg albumen after the shell is removed.

More recently, government agents during the Second World War photographically reduced text to a dot less than one millimeter in diameter. Agents often concealed their microdots in plain sight: sitting on top of a period in an apparently innocuous letter. In the 1950s, a Soviet agent shrouded his messages in hollow cufflinks and nickels. Today, a digitized image may conceal a document, a map or a spreadsheet.

The ruse of camouflaging a message has a flaw: Upon discovery, anyone can read the contents. Cryptography, the technique of concealing a message's meaning, developed with steganography. Over the centuries, cryptographers devised new ways to scramble a message, and invented techniques to unscramble the messages of enemies. While the makers and breakers of secret writing competed, they advanced mathematics, linguistics and information theory, as well as laid the foundation for computer science. They also made history.

Conflicts Nurture Secret Writing

Cryptography works like this. Two parties who wish to communicate in secret agree upon a method to conceal a message's meaning. The sender scrambles a message using this modification protocol, and the recipient uses the protocol to unscramble the message. A third party may intercept the message, but cannot read it without knowing the key that determines the modification protocol.

A message can be scrambled, or encrypted, using two basic methods. A cryptographer may use transposition to rearrange the message's letters and create something akin to an anagram, or a cryptographer may substitute the message's letters with other letters or symbols. Altering a message at the level of its letters produces an enciphered text. Substituting a message's words or phrases for words, numbers or syllables produces an encoded text.

The first military cryptographic device transposed messages. In the 5th century BC, the Greek historian Thucydides credited the Spartans with developing the scytale, a wooden staff around which the message sender wound a strip of cloth, papyrus, or leather. The sender wrote the message along the length of the scytale, and then unwound the strip, which bore the letters in nonsequential order. To recover the message, the receiver wrapped the strip around a wooden stick of the same diameter as the sender's scytale.

Julius Caesar often used secret writing, including substitution ciphers. In one type of cipher, he rearranged the alphabet with a simple shift substitution, now known as the Caesar cipher. The message sender enciphers the text by replacing each letter with a letter found at a fixed number of places downstream in the alphabet. A shift of three, for example, transforms "cat" to "fdw."

Between 800 and 1200 AD, cryptography advanced in the complex, organized Islamic civilization. During this time, administrators encrypted messages about sensitive affairs of state and even tax records. They used cipher alphabets containing a rearrangement of the plain alphabet, as well as cipher alphabets that contained other symbols.

The one-to-one substitutions of such monoalphabetic substitution ciphers can be cracked. The 9th century Baghdad scientist, al-Kindi, described a method for unscrambling a message – altered using a simple substitution cipher – without knowing the process used to scramble it. In short, he described a technique of cryptanalysis.

A language has frequency patterns for vowels, consonants and syllable pairs. When letters of the plain alphabet are replaced with symbols to create a cipher, the symbols take on the characteristics of the original letters, such as relative frequencies. In the English language, for example, the letter E accounts for 13 percent of all letters. When a symbol replaces E, that symbol will account for 13 percent of all symbols in the enciphered text. A cryptanalyst uses these letter frequencies to decipher a message.

While cryptography matured in the Islamic world, Europe struggled with its Dark Ages. Medieval monks, fascinated by the Old Testament's substitution ciphers, helped to reintroduce encryption into Western civilization. In the 13th century, English Franciscan monk Roger Bacon wrote the first known European book to describe the use of cryptography.

By the 15th century, cryptography had become widespread in Europe. The Renaissance revival in the arts and sciences nurtured cryptographic techniques. But it was politics that fueled the advancement of secret message making and breaking, especially the politics that broiled among feuding leaders of the Italian city-states. Giovanni Soro, one of the great European cryptanalysts of the early 16th century, served as Venetian cipher secretary. Soro decrypted messages for Venice, friendly city-states and Pope Clement VII.

With the increased use of cryptography, experts became aware that frequency analysis could defeat simple monoalphabetic substitution ciphers. To complicate frequency analysis, cipher writers misspelled words before encrypting the message, or they included nulls – symbols or letters that represented nothing.

The use of codewords offered a way to evade a cryptanalyst's frequency analysis. A cryptologist encodes a message by substituting words or phrases, which are much less vulnerable to frequency analysis. Codes, however, have several significant drawbacks of their own.

To use a cipher, a sender and receiver must agree upon the letters of the cipher alphabet. Once they have done so, they can encipher and decipher any message. To achieve this level of flexibility with coded messages, a sender and receiver must devise a code word for every plaintext word that they plan to use. After arduously compiling a codebook, the sender and receiver must tote it around. If the enemy obtains a copy of the codebook, then the enemy can read all encoded communications. Senders and receivers must compile a new codebook.

In the 16th century, cryptographers realized the weakness of codes and turned to nomenclators, which combined cipher alphabets and codes for words, syllables and names. Nomenclators persisted as a popular masking system for 300 years.

The nomenclator system was not foolproof. During the violent conflict between Catholics and Protestants in 16th century England, Queen Elizabeth I placed Mary, Queen of Scots, in virtual imprisonment. Here, she became a symbol of Catholic oppression. Using a nomenclator, Mary cautiously agreed to a proposal to seize the throne from Elizabeth. After Thomas Phelippes deciphered the letter, the Queen ordered Mary's execution.

A new tactic for secret writing simmered among cryptographers. In the 15th century, Florentine architect Leon Battista Alberti proposed an encryption system in which the writer switched back and forth between two or more cipher substitution alphabets. Over the next century, polyalphabetic ciphers developed in the hands of Johannes Trithemius, a German abbot, Giovanni Porta, an Italian scientist and Blaise de Vigènere, a French diplomat. The Vigènere cipher uses 26 cipher alphabets to encrypt a message. Although the Vigènere cipher defeats frequency analysis, cryptographers made little use of it for centuries.

By the 1700s, heads of state rounded up cryptanalysts, who attacked monoalphabetic ciphers. European governments formed centers for decrypting messages and gathering intelligence, the Black Chambers. England had a Deciphering Branch; France, the Cabinet Noir; Russia had the secret police; and Austria, the Geheime Kabinsets-Kanzlei. As each side cracked foreign secret writing and improved their own secret writing, they developed new ciphers and codes.

Possibly the oldest existing cipher instrument, this 18th century device has scrambled alphabets on the edge of each wheel for enciphering a message. Plaintext letters are aligned in one row and other rows can be chosen for the cipher text. Source: US National Security Agency.

Cryptography Finds a New Home in America

The practice of secret writing in military conflicts traveled to North America. During the American Revolution, the British, their Loyalist allies and the Colonials used a variety of secret writing systems, including codes, ciphers and nomenclators.

One of the most efficient Colonial cryptology groups operated in Loyalist-dominated New York City. From here, spymasters communicated with General Washington using a nomenclator and messages concealed with stains or invisible inks. They also used a dictionary-based code, in which numbers indicated the locations of intended words in an edition of a particular dictionary known to the sender and receiver. Secret writing allowed rebel agents to send valuable information about British troop strength, morale and supplies.

In 1838, Samuel Morse transmitted the first recorded telegraph message. The development of Morse code showed that numbers and letters can be transformed into dot-dash patterns. Both sides relied upon the telegraph during the American Civil War. Abraham Lincoln in particular used the new technology to stay in touch with his forces almost in real time. But other, more secret, means of communication also played important roles.

Union forces employed transposition ciphers to conceal telegraph messages from Confederate wiretappers and cryptanalysts. Anson Stager, Western Union's first superintendent, devised a word transposition cipher used by Union General George B. McClellan. Historians credit Rose Greenhow with leaking strategic information about Union troop movements and numbers that helped Confederate General Pierre G.T. Beauregard achieve victory in the 1861 Battle of Bull Run. Greenhow transmitted the information with a number and symbol substitution cipher.

Secret Writing in World Wars, Cold War and Cyberspace

With the increasing unrest in Europe that presaged the First World War, England's Parliament ordered an increase in the military's efforts to intercept and decrypt messages. In 1914, this resulted in the birth of a cryptanalytic group known as Section 25 of the Intelligence Division, or "Room 40." It became the Allies' leading code and cipher branch.

Shortly after its creation, Room 40 received a gift: a German codebook obtained by the Russian Navy from a German light cruiser. A British trawler recovered another German codebook. Within weeks, the codebreakers began to decode radio intercepts. Room 40's efforts led to the discovery of a plan for German warships to converge at Dogger Bank, located off

England's northern coast. The British navy met the German vessels, severely damaging two warships and destroying a third.

One of the most famous decryption efforts revealed the contents of an encoded telegram sent on January 16, 1917, by German Foreign Minister Arthur Zimmermann to the German Minister in Mexico. Germany planned to begin unrestricted submarine warfare against neutral shipping, a strategy that could compel the United States to join the Allies. Concerned that the United States might not stay neutral, Zimmermann outlined a proposal for an alliance with Mexico: "make war together, make peace together, generous financial support and an understanding on our part that Mexico is to reconquer the lost territory in Texas, New Mexico, and Arizona." Room 40 intercepted the telegram and decrypted the contents. The revelation spurred the United States into the war effort.

After the war, ciphers surpassed codes as a military concealment method. Advancements in electromechanical technology allowed the production of rotor machines that quickly reordered letters to create polyalphabetic ciphers. In the late 1920s, the German military modified an encrypting machine built for use in business. It was called the Enigma and could create immense numbers of electrically-generated alphabets.

In the early months of the Second World War, Britain's new prime minister, Winston Churchill, supported a cryptanalysis group at Bletchley Park. Aided by Polish and French sources, Alan Turing led the British effort to crack the Enigma. By late April 1940, Bletchley Park had a machine based on the Poles' "Bombe." The electrical circuits of the eight foot high machine imitated the Enigma's rotors. The Bombe decrypted Enigma transmissions recorded by England's intercept stations. Information gleaned from decrypted Nazi messages helped the Allies to win the war.

During the Second World War, the US Navy, concerned about shipping losses, asked engineers at the National Cash Register Company to redesign the Bombe, a machine invented by Polish and English engineers to read messages enciphered on the German Enigma. From 1943 to 1945, 121 Bombes were installed at the Navy's Communications Annex in Washington D.C. WAVES (Women Accepted for Volunteer Emergency Service) operated the Bombes 24 hours a day, 7 days a week. The efforts played a vital role in ending the war in the Atlantic and Europe. Source: US National Security Agency.

As secret writing flourished in the Cold War, civilian uses for encryption also grew. Banks needed to secure their transactions and healthcare providers needed to protect confidential patient data. Today, rigorous data privacy laws encourage the private sector to embrace encryption technology. Internet commerce depends upon strong encryption to ensure the privacy and security of transactions. At the same time, law enforcement agencies contend with criminals who encrypt evidence and terrorists who conceal their plans with encryption. Secret writing remains intertwined with conflicts.

Code Talkers

The US military did not abandon code concealments during the Second World War. The Marine Corps recruited Native Americans who communicated in their language. These code talkers overcame the basic flaw of codes: They did not need codebooks.

Philip Johnston, the son of a Protestant missionary to the Navajos, was a veteran of the First World War, who knew that the military had used Native American languages, such as Choctaw, to send secret messages. Johnston thought that the Navajo language might provide an unbreakable code for the current war. This unwritten language is extremely complex, has no alphabet or symbols, and relies upon complicated syntax and a speaker's tonal variations. Extensive exposure to the language and training are required to understand it.

During May 1942, Johnston's first 29 Navajo recruits arrived in a California boot camp. After the war, officials discovered that the recruits who passed Camp Pendleton basic training varied in age from 15 to 35. The group called "the original 29" devised a Navajo code suitable for military purposes, a code that code talkers had to memorize.

The first code had 211 words. Most were Navajo words that had been given a new military meaning. A "fighter plane," for example, was called "da-he-tih-hi," which means "hummingbird" in Navajo. The Navajo word for "chicken hawk" ("gini") meant "dive bomber." "Besh- lo" (iron fish) meant "submarine," whereas the word for "tortoise" ("chay-da-gahi") meant "tank." In addition to these code words, the code talkers spelled out words that were not in the Navajo language vocabulary. They achieved this by combining a group of apparently unrelated Navajo words. To decode the message, the code talker translated each Navajo word into its English equivalent. Once he had a list of English equivalents, he used only the first letter of each word to spell out the encoded word. For example, the word "navy" in Navajo code would be "tsah" (needle) "wol-la-chee" (ant) "ah-keh-di-glini" (victor) "tsah-ah-dzoh" (yucca).

An expert in secret writing studied the code talkers' transmissions and became concerned that frequency analysis might reveal the contents of a message when code talkers used the alphabet to spell out words foreign to the Navajos' vocabulary. So, the code talker alphabet was revised. They increased the number of words used for the most frequently repeated letters. In the new code, the letter A, for instance, could be designated by the Navajo words "wol-la-chee" (ant), "be-la-sana" (apple), or "tse-nill" (axe).

The Marines assigned Navaho code talkers to military units based in the Pacific battles. Here, the code talkers transmitted information on strategies, troop movements, troop strengths, and battlefield orders. Major Howard Connor, 5th Marine Division signal officer, had six Navajo code talkers in his group. During the first two days of battle at Iwo Jima, they sent and received more than 800 messages – without making a mistake.

"Were it not for the Navajos," Connor said, "the Marines would never have taken Iwo Jima."

About 400 Navajos served as code talkers. The code talker program was highly classified throughout the war and remained a secret until 1968. Only recently have the code talkers earned widespread recognition from the government and the public. In 1992, 35 US Marine code talkers were honored for their contributions at a ceremony held in the Pentagon, Washington, D.C. In 2001, President Bush presented Congressional Gold Medals to the first 29 Marine code talkers.

With the Code Talkers Recognition Act of 2008, the US government finally acknowledged the role of Comanche code talkers in both World Wars. In the Second Word War, the US Army had 14 Comanche code talkers in the European theater. The French Government had awarded the group's three survivors the "Chevalier de L'Order National du Merite" in 1989.

Copy Machines

In 1780, the prolific Scottish inventor James Watt patented a letterpress that produced copies of handwritten letters. Using rollers or a screw mechanism, an office clerk pressed a thin piece of paper against text written in special ink. The reversed copy of the writing could be read through the back of the thin paper.

While the letterpress enjoyed widespread use during the 19th century, other inventors devised their own solutions to copy making. One approach required the use of corrosive ink or a tooth-edged pen to create a stencil template for making inked copies.

During the 1870s, Eugenio de Zuccato and Thomas Edison invented stencil-based duplicating processes. Albert B. Dick licensed the Edison patents and sold a mimeograph that used handwritten stencils made with waxed paper. By the late 1880s, the mimeograph churned out copies from typewritten stencils. The technology had a serious limitation: Duplication required a stencil.

With photocopying, any document could be duplicated. The first commercially important photocopying technique used a cyanotype process invented by astronomer John Herschel in 1842. Hershel had discovered that sunlight transforms colorless, water-soluble iron salts to a compound called Prussian Blue. In the early 1870s, European clerks used photosensitive paper to make blue-colored negative images – blueprints – from architectural drawings. By the early 20th century, the introduction of new chemical mixtures enabled the creation of copies with white backgrounds.

In the midst of America's Great Depression, Charles F. Carlson analyzed patents for an electronics firm. Frustrated by the need to make copies of patents, Carlson experimented in his kitchen and invented a copying process based upon the interaction between light and electrostatic fields. He hired an unemployed physicist, Otto Kornei, and rented a room above a bar. Here, Carlson and Kornei developed a technique that would be called "xerography," based upon the Greek for "dry writing."

On October 22, 1938, they coated a zinc plate with sulfur and rubbed it with a handkerchief to impart an electrostatic charge. Using India ink, Kornei wrote the date and location of the lab on a glass microscope slide, placed the slide on the plate, and exposed it to a bright incandescent lamp. Light increased conductivity, causing exposed sulfur to lose its charge. The sulfur coat retained its charge in areas masked by the writing on the slide. They dusted the plate with moss spores, which clung to charged areas, and blew away loose spores to reveal writing. To make a permanent copy, they transferred spores to wax paper and heated it to melt the wax. In time, an electrostatically-charged rotating drum replaced the zinc-sulfur sheet, and dry ink replaced moss spores.

After a long search for a financial backer, Carlson struck a deal with Battelle Memorial Institute in 1944. Three years later, Battelle secured additional funding from Haloid, a small photographic technology company.

Meanwhile, Carl Miller at the Minnesota Mining and Manufacturing Company invented a copying process in which he bombarded infrared light on a sheet of heat-sensitive paper that lay on top of an original document with carbon-based ink. Carbon on the original document absorbed infrared energy, became warm, and transferred heat to the heat-sensitive paper to produce a blackened copy.

3M brought out the Thermo-Fax copier in 1951, while Haloid Company, now calling itself Xerox Corporation, introduced its xerography-based office copier in 1959. The two technologies competed through the next decade, each shackled by drawbacks. Thermo-fax technology produced copies with ink that faded and paper that became brittle. Although Xerox technology used plain paper, its early machines proved cumbersome. In time, Xerox improved its machines to become the dominant office copying technology.

Monastic Sign Language

Within a monastery, the Rule of St. Benedict, circa 530, forbade conversation during certain periods of the day and in certain places. Silence could be broken, if necessary, by some sign other than speech, an exception that encouraged the development of hand languages.

Codifications of Benedictine hand signs date to the late 11th century at the Abbey of Cluny in present day France and the Hirsau Abbey in present day Germany. The gestures were not intended to replace language, but simply served as a means to communicate without breaking a period of silence. Monks' sign lexicons represent a type of symbolic gestural communication. In one form of the Cluny sign lexicon, for example, a monk would indicate a fish by moving his hand to imitate the motion of a fish's tail.

European monks also developed systems of alphabetic gestures, known as finger spelling. In the 7th century, Saint Bede the Venerable, an Anglo-Saxon Benedictine monk, described a system for representing the alphabet using fingers, possibly as a parlor game. Eventually, monks used manual alphabets as mnemonic devices, as shown by Franciscan friar Cosmas Rosselius' *Thesaurus Artificiosæ Memoriæ* (*A Treasury of Artificial Memory Techniques*) [1579].

One of Rosellius' one-hand alphabets played an important role: Spanish Benedictine monk Pedro Ponce de León adapted the system to teach the deaf. In 1578, Ponce de León described how he taught congenitally deaf sons of the wealthy to read, write, study science and history, and to pray. Susan Plann, author of *A Silent Minority: Deaf Education in Spain, 1550-1835* (1997), traced the dispersion of manual alphabets in European deaf communities from the efforts of 16th and 17th century monks to tutor deaf children.

V. Civil Engineering

Sub Rosa Subway

By the mid-19th century, Manhattan experienced an untidy gridlock. As a rapidly growing population squeezed into the narrow island, pedestrians and horse-drawn conveyances jammed the streets. Every horse potentially donated as much as ten pounds of manure to the roads daily.

Alfred Ely Beach, like many of his contemporaries, believed in technological fixes. He envisioned a way to relieve his city's transportation problem: travel underground. Beach thought that an air pressure-driven subway would provide a solution, a larger version of pneumatic tubes that delivered mail and small packages in London.

In 1868, Beach took an indirect route to realize his pneumatic subway. He sought the government's permission to build an experimental pneumatic package delivery system of two tunnels – each less than four and one half feet in diameter. The state legislature granted the charter. Then, Beach requested and received an amendment to allow construction of a large tunnel that would contain two smaller tubes.

Historians suggest that Beach had first requested permission for two small tunnels to evade the notice of those who ran New York City: William M. Tweed and Tammany Hall. Boss Tweed and his group collected fees from owners of private streetcar lines and planned to expand operations with an elevated railroad.

The Beach Pneumatic Transit Company leased the basement under Devlin & Co., located across the street from City Hall. After enlarging the basement, workers cut a tunnel with a hydraulic shield, originally designed by French-born British engineer Marc Brunel and modified by Beach. Hydraulic rams drove the barrel-shaped device, which ground the earth. At night, wagons carted away bags of soil.

By January 1870, Beach's crew had constructed a 300-foot brick-lined tunnel that started at Warren Street and Broadway and ran under Broadway to Murray Street. The true purpose of the passage escaped notice. The *New York Times* reported that the tunnel had an interior diameter of eight feet, which would be divided to form a double line of pneumatic tubes.

Amidst unfounded speculation that the tunnel undermined the streets above, the city government attempted to halt construction. To raise support for his pneumatic subway, Beach opened the unauthorized railway on February 26. Visitors were shocked.

"Such as expected to find a dismal, cavernous retreat under Broadway," reported the *New York Times*, "opened their eyes at the elegant reception room, the light, airy tunnel and the general appearance of taste and comfort in all the apartments." The subway's depot boasted a water fountain, chandeliers, paintings and a grand piano.

A cylindrical car shuttled visitors between stations. Illuminated with zircon lamps, the car had oval windows at each end, plush-lined seats, and floors of oiled wood.

Air pressure drove the car. A steam engine powered a blower that could deliver 100,000 cubic feet of air per minute. The artificial gale blew the car at a speed of ten miles an hour. To slow the car, operators reversed the blower when the car reached the midway point. Switching the direction of air flow also produced a partial vacuum that allowed atmospheric pressure to propel the car back to the first station.

Throngs of enthusiastic visitors paid 25 cents for a ride on the subway, a sum that Beach donated to charity. Within the first two weeks, sales totaled almost $3,000.

Interior of the passenger car. Source: *Frank Leslie's Illustrated Newspaper* 29:381 (February 19, 1870).

Encouraged by the public's response, Beach campaigned for an amendment of his company's charter to extend the subway uptown for about five miles. The state legislature eventually passed it. During the same session, they also approved a Tammany-backed bill for an elevated railroad. On April 1, 1871, Governor John T. Hoffman vetoed the Beach Pneumatic Railway bill.

Beach did not give up; he continued to lobby for his subway. In 1873, he tried for the fourth time. By now, Boss Tweed had been removed from power. The legislature approved the bill, and Governor John A. Dix signed the bill into law.

Before he could start on the railway, the stock market collapsed. Beach's company already experienced trouble in attracting support. After the Panic of 1873, investors disappeared. The tunnel served as a shooting gallery, a storage vault, and then it was sealed.

In 1904, the city opened its primary subway system, one powered by electricity. Eight years later, workers broke into Beach's tunnel while excavating a new branch of the Brooklyn-Manhattan Transit. Here, they found a deteriorated rail car sitting on corroded tracks, a dried water fountain and a waiting room embellished with décor from the previous century. They had uncovered America's first subway.

The Statue of Liberty Enlightening the World

In the summer of 1865, French intellectuals gathered for a dinner party near Versailles. The diners, who opposed Napoleon III's oppressive rule, discussed the American success in establishing a democracy, a success that owed a debt to the generous support of France during the American Revolution. The scholar Édouard René Lefèvre de Laboulaye noted that, in eleven years, Americans would celebrate the centennial of the country's independence. He suggested that the French and Americans should collaborate once again; this time, to build a monument to commemorate American independence.

Laboulaye's suggestion made a lasting impression on at least one of the guests: Frédéric-Auguste Bartholdi, a successful 31-year-old sculptor. "[T]his conversation interested me so deeply that it became fixed in my memory," Bartholdi said in a book written years later. He wrote *The Statue of Liberty Enlightening the World* (1885) to encourage donations from the US public to build the statue's pedestal; the US government had shown little interest in the project.

The Statue Enjoys an Extended Stay in Paris

After fighting in the Franco-Prussian War and watching the rise of France's Third Republic, Bartholdi met again with Laboulaye, who suggested that the sculptor should visit the United States. "Go to see that country," he said. "You will study it, you will bring back to us your impressions." Laboulaye told Bartholdi to promote the idea of a monument to honor the friendship between France and the United States. The project, he predicted, would have a far-reaching moral effect in both countries.

Bartholdi took his friend's advice. In 1871, as his ship entered New York Harbor, the sculptor saw the perfect location for the monument: Bedloe's

Island. Positioned at the portal of the country, the land happened to be owned by the US government. While Bartholdi traveled across America, he advanced his plan with sketches of Lady Liberty. His proposal was met with interest, but it failed to excite monetary commitments.

Following his return to France, Bartholdi focused on other projects. One of these, a statue of Marquis de Lafayette cast in bronze, would be presented to New York City as a gift from France. In 1874, Bartholdi and Laboulaye reconsidered a French-American monument. They decided that France could pay for the statue, and that the United States could pay for the pedestal and foundation.

The necessity of shipping the statue overseas limited the material used in the monument; stone and bronze would be too heavy. Bartholdi worked with Caget, Gauthier and Company, a group of experts in repoussé, a metalworking technique for creating a sculptural form by hammering malleable metal from the reverse side.

To ensure the accuracy of the colossal statue's proportions, Bartholdi made a series of progressively larger models. First, he made a statue four feet high, which was used to cast a model about nine feet high. He reproduced this model as a 36 foot plaster statue supported by a wooden frame. After correcting the model's details, the statue was divided into nearly 300 sections, which would be reproduced about four times in size. Nine thousand measurements were required to replicate a single section of the final statue.

Carpenters fitted a mold of laminated wood against the plaster surface of each of the enlarged, final sections. Workers used mallets and rammers to press copper sheets against the mold. Copper rivets joined the shaped copper pieces together for the envelope of the statue. This cover would be supported with a skeleton of iron and steel, designed by Alexandre-Gustave Eiffel, who would later build Paris' famous tower.

In the beginning, Bartholdi and Laboulaye hoped to honor America's centennial by presenting the Statue of Liberty to the United States on July 4, 1876. However, they could not raise sufficient money to meet that date. Bartholdi did have the chance to display a 30-foot raised arm and torch on the final days of the 1876 International Centennial Exhibition in Philadelphia. Visitors climbed a ladder to a balcony that surrounded Lady Liberty's torch.

The torch and part of the arm of the Statue of Liberty on display at the 1876 Centennial Exhibition in Philadelphia. Source: US Library of Congress.

The Franco-American Union, which had been struggling to raise funds for the statue, tried a new strategy: They held a lottery. Prizes included a silver plate valued at about US$20,000, expensive jewelry, and two works by Bartholdi. The group also sold numbered and signed clay models of the statute. By July 1882, the group had raised adequate funds to complete Lady Liberty, a project that now ran about US$250,000. All of the money had come from individual donations.

Bartholdi and his craftsmen completed the statue in 1883. Since the US pedestal had yet to be built, the statue remained in Paris' Rue de Chazelles. Lady Liberty would enjoy the view far longer than Bartholdi anticipated.

Statue and Pedestal Convene

In 1877, the US Congress authorized the setting aside of ground in New York Harbor for the monument. As Bartholdi had requested, the monument would reside on Bedloe's Island. Under the guidance of chief engineer General Charles P. Stone, the pedestal would be built in the center of Fort Wood, an eleven-pointed, star-shaped battery constructed in 1811 to defend the harbor. The pedestal's granite outer wall would encase a massive shaft of concrete reinforced with girders. Designed by Richard Morris Hunt, the pedestal would be one of the heaviest pieces of masonry ever built.

Funds for the pedestal trickled in. Congress rejected a bill that would have appropriated $100,000 for the base. The New York legislature approved a grant of $50,000, but the governor vetoed it. Individuals were reluctant to donate until they could be convinced that the French would complete the enormous statue. Outside of New York, many Americans decided that the city itself could pay for "New York's lighthouse."

Efforts by the American half of the Franco-American Union met apathy. By 1884, the organization had collected $182,491, and most of this had been spent. Work on the pedestal stopped in the fall; less than 20 percent had been completed. The American committee needed to raise an additional $100,000.

When Joseph Pulitzer learned that a lack of funds might doom the Statue of Liberty, he decided to use his two newspapers, the *New York World* and the *St. Louis Post-Dispatch*, to raise money, as well as circulation. In the financial newspaper, the *World*, he blasted the wealthy for their selfishness, while promising to publish the name of anyone who contributed to the monument project. Pulitzer stressed that the money for the pedestal should be donated by all Americans, not by the wealthy and not by the government. After all, the statue had been financed by the people of France. Representatives from Cleveland, Boston, Minneapolis, Philadelphia, San Francisco and other cities offered to pay the costs of the pedestal. But the generous offers had a condition: The statue had to be erected in the city that donated the money.

Part of the statue displayed in a Paris park (1883). Source: US Library of Congress.

In 1885, Bartholdi ordered the Statue of Liberty dismantled to 350 pieces, and packed in 214 wooden crates, weighing from a few hundred pounds to several tons each. The shipment arrived at Bedloe's Island on June 15, 1885. Two months later, the *World* announced that more than 120,000 individuals had donated to the project. They had collected a little over $100,000.

The Statue of Liberty was dedicated on a misty, cold October 28, 1886. More than a million people lined the streets to watch a grand military and civic parade that included Bartholdi and his wife. President Grover Cleveland boarded the *U.S.S. Despatch* along with other participants in the ceremony on Bedloe's Island. After leading a double column of about 300 tugs, yachts and excursion steamers, the *Despatch* anchored at the island alongside US and French warships. Vessels that had followed the *Despatch* carried their audiences to assigned positions around the island. Accidentally terminating the oration of a US Senator, Bartholdi tugged on a rope. A rain-soaked French tricolor covering fell from Lady Liberty's face, literally unveiling the statue.

Apparently the event did not please everyone. On November 1, the *New York Times* published a report from Paris that needled the English, who were "furious" over the goodwill evidenced by the Statue of Liberty. "France laughs at their fury," the reporter said, "which does not in the least interfere with the general satisfaction."

Statuesque Facts

The statute itself stands 151 feet, 1 inch high.

The statue rests on a pedestal and base 154 feet from the ground.

For 13 years, the Statue of Liberty was the tallest structure in New York.

The statue contains 100 tons of copper and 125 tons of steel.

The concrete foundation weighs 27,000 tons.

Folds of the statue's drapery allow copper sheets to expand or contract with changing temperature.

The pedestal is anchored so securely to the foundation and the rock below that a wind storm would almost have to capsize the island before overturning the statue.

The crown's 25 windows symbolize gemstones found on the earth and the heaven's light shining over the world.

The crown's seven rays represent the seven seas and continents of the world.

Charlotte Beysser Bartholdi, the sculptor's mother, was the model for the Statue of Liberty.

Reinforced Concrete

For more than seven thousand years, humans have transformed the environment with concrete, a mixture of cement, water, sand and bits of stone. The Egyptians became the first substantial concrete users around 2,500 BC. Two thousand years later, Roman engineers eagerly favored concrete structures over traditional Greek construction of wood and cut stone. The Romans imprinted their empire with concrete foundations, terraces, arches, domes and vaults.

One of the most durable and versatile of building materials, concrete has faults. Although concrete effectively resists compressive stresses, tensile forces – stretching or bending – break concrete's rigid lattice resulting in cracking and rupture. Concrete can be reinforced by embedding an elastic material, such as metal. This reinforced concrete can resist stretching, bending and other tensile forces.

During the mid-19th century, Monsieur J.L. Lambot of Provence, France, demonstrated the usefulness of reinforcing by building boats with mortar reinforced with wire mesh and iron bars. He also produced reinforced concrete planters. Joseph Monier, a gardener at the Versailles Palace, developed his own process for making sturdy concrete garden tubs and flowerpots formed around wire mesh. Monier patented his method and licensed the patent. One of Monier's licensees, German engineer, G.A. Wayss, adapted the technique for the construction of bridges.

In 1854, William B. Wilkinson of Newcastle, England, introduced reinforced concrete in the building of houses. He built a two-story servant's cottage, in which he reinforced the concrete floor and roof with iron bars. Twenty years later, William E. Ward used iron bars for reinforcing concrete in a Port Chester, New York house. Ward's Castle marked the first reinforced concrete building in the United States.

During the 1890s, Austrian engineers advanced the theory and practice of embedding steel reinforcing bars, or rebar, within concrete mixtures. French engineer, François Hennebique, popularized concrete reinforced with straight round steel rods. In 1897, he built the Weaver's Flower Mill (Swansea, Wales), credited as Britain's first reinforced concrete building.

While Hennebique promoted reinforced concrete in Europe, Ernest L. Ransome developed new methods for using the material in North America, including a system of twisting square steel bars to increase tensile strength and adhesion to concrete. He also devised methods for casting reinforced columns, girders, stairways, wall panels and other elements at a building site. In 1889, Ransome built one of America's first reinforced concrete bridges: the Alvord Lake Bridge in Golden Gate Park (San Francisco, California).

At the dawn of the new century, the use of reinforced concrete spread through North America, Europe and Australia. The structural material supported houses, apartment buildings, warehouses, factory buildings and bridges.

The year 1903 saw the completion of the first reinforced concrete skyscraper: the Ingalls Building in Cincinnati, Ohio. Skeptics predicted that the 16-story high-rise would rapidly crumble under its own weight. When the building stood intact, the success ushered in an era of skyscrapers, forever altering cityscapes. In the decades ahead, however, steel construction, not reinforced concrete, dominated the building of skyscrapers.

Meanwhile, architects employed reinforced concrete to move away from traditional techniques that imitated wood and to realize artistic expression in new forms. These efforts can be seen in Frank Lloyd Wright's Solomon R. Guggenheim Museum in New York City, Le Corbusier's Chapel of Nôtre Dame du Haut in Ronchamp, France, and other sculptured buildings.

By the 1960's, advancements in formwork techniques, high-strength concrete formulations, and reinforcement methods ensured a primary position for reinforced concrete in high-rises. From Chicago's 179-meter tall Marina City Complex (1964), to the 452-meter tall Petronas Towers (1998) of Kuala Lumpur, Malaysia, to Dubai's 828-meter tall Burj Dubai, reinforced concrete has brought a new meaning to the word "skyscraper."

The Air Brake: Stopping a Train with Wind

While cruising at 15 miles per hour on Canada's Grand Trunk Railway, a train entered the Beloeil Bridge in the early hours of June 29, 1864. Engineer William Burney threw his locomotive into reverse, attempting to stop before he reached the open drawbridge. But it was too late. The train plunged through the gap toward the Richelieu River and hit a barge passing below. Ten passenger cars piled one upon another, "crushed into an unrecognizable mass of splinters and iron," according to the *Montreal Gazette*. Ninety-nine people died.

North American railway travel changed during the second half of the 19th century. New, powerful locomotives enabled trains to lengthen and move at greater speeds. These developments created a challenge: stopping the trains.

Reducing a train's speed with mechanical brakes was dangerous and unreliable. When the engine sounded the "down breaks" whistle, brakemen ran from car to car – sometimes along the roofs – tightening brake wheels with pick handles. Since the process could not be coordinated in any practical way, passengers experienced continual discomfort and trains tended to crash.

In 1866, two freight trains collided in New York, delaying a passenger train that carried 20-year old George Westinghouse. The collision could have been avoided, Westinghouse decided, if the trains had been equipped with an efficient brake system. Ideally, train engineers should be able to operate brakes on all cars simultaneously. How could this be accomplished?

Following a three-month college career, Westinghouse worked at his father's factory, which produced agriculture machinery. It was here that Westinghouse turned his attention to the problem of brakes. At first, he tried to modify a system that applied brakes to all cars via a chain that ran the length of the train. Insurmountable mechanical problems ended this effort.

By chance, two young women had convinced Westinghouse to purchase a magazine subscription. One of the first issues included an article about the construction of the Mont Cenis railroad tunnel project, an effort to connect towns in France and Italy across the Alps. At both headings of the tunnel, workers bore through rock with compressed air-driven drills. Westinghouse realized that compressed air could transmit power from a train's locomotive to brake mechanisms on each car.

In 1869, Westinghouse obtained a patent for his straight-air brake. The brake system relied upon a locomotive's steam engine to compress air into a reservoir linked by pipe to the engineer's brake valve. The valve controlled the flow of air into a line of pipes and flexible connectors that passed under all of the cars, ultimately joining with pistons connected to brake shoes. An engineer released the brakes by moving the handle of the brake valve to cut off communication with the reservoir and bleed air from the brake pipe to the atmosphere.

Westinghouse's new brake system made history with its trial run on a Panhandle Railroad accommodation train. As the train emerged from a tunnel near Pittsburgh's Union Station, the engineer saw a farmer's wagon stalled on the tracks. Thanks to the air brake, the engineer halted the train in time to prevent a collision. Within a year, the young inventor established the Westinghouse Air Brake Manufacturing Company in Hamilton, Ontario. He would later relocate his company to a facility with more than nine acres of floor space in Wilmerding, Pennsylvania.

The straight-air brake had a few flaws. Although faster than a mechanical brake, compressed air still took too much time to reach the end of the train. In addition, a rupture in the hose connection between two cars would disable the brakes.

In 1871, Westinghouse began a series of improvements that led to an industry standard. His major innovation was to reverse the operation of the air brake system. Now, filling the brake pipe with air charged the system and released the brakes, whereas draining air from pipes applied the brakes. Not only was Westinghouse's automatic brake more responsive, but it had a safety feature: The brakes activated if the connection between two cars disrupted.

The air brake's promise of safe travel at high speeds revolutionized the railroad industry. By the end of the 19th century, US law required the use of air brakes on all interstate rail lines.

Despite the benefits offered by the air brake, some railroad men had initially greeted the device with skepticism. When Westinghouse approached Commodore Cornelius Vanderbilt with his invention, the American railroad magnate dismissed him. "If I understand you, young man," Vanderbilt said, "you propose to stop a railroad train with wind." That's exactly what Westinghouse achieved.

Engineering Feats of James Eads

In September 1833, thirteen-year-old James Buchanan Eads, his mother and two sisters traversed the Mississippi River on a riverboat. They planned to set up a new home in St. Louis, but as they approached the dock, a chimney flue collapsed. Eads and his family barely survived the ensuing fire. Their possessions went up in flames. Eads abandoned formal education and worked to help support his family. In time, he became an errand boy for Williams & Duhring dry goods store, where Eads used a small, private library to educate himself about physical science, mechanics, machinery and civil engineering.

At the age of nineteen, Eads returned to the Mississippi River as a clerk on the steamboat *Knickerbocker*. In 1840, while laden with a large cargo of lead, the steamboat was sunk by an underwater snag near Cairo, Illinois. The loss of a boat and cargo was not unusual; wrecked boats littered the riverbed. A fortune lay buried under the river's relentlessly shifting sand.

Eads decided to recover that fortune. By 1841, he had designed a diving bell that allowed a diver to walk on the bottom of the Mississippi. Eads and two boat builders established a thriving salvage business. Eads' experience gave him a practical education in the Mississippi River's currents and the behavior of the riverbed's sand and silt. The salvage business also made Eads a rich man.

Soon after the American Civil War erupted in April 1861, Attorney General Edward Bates requested Eads to attend a conference about the use of gunboats. Eads suggested that Union soldiers could use shallow-draft gunboats to take Confederate forts. He won a bid to build seven 500 ton, 175-foot long armored wooden gunboats, the City Class Ironclads, which formed a unique fleet designed for the country's inland waterways. During the war,

Eads also invented a steam-driven turret used on river monitors. Eads' contributions played an integral role in the successful Union campaign to divide the Confederacy by opening the Mississippi River from Cairo to New Orleans, and in Admiral David Glasgow Farragut's victory at Mobile Bay.

Patterns of commerce shifted radically after the Civil War. Chicago transformed into a commercial center as railroads became the country's main movers of trade. Shippers tried to avoid routing their cargo through St. Louis, because it required off-loading from trains to boats to trains once again in that city. To regain its dominance, St. Louis needed a railroad bridge over the Mississippi River.

In 1867, Eads proposed a bridge with two decks: an upper level for carriages and horses and a lower deck for the railroad, linking St. Louis to the rail lines that ran east to west. Three massive cantilevered arches – one arch 520 feet long and two arches of 502 feet – would support the flow of this traffic above the Mississippi. The bridge would have a span longer than any bridge ever built and its supporting piers would stand on bedrock. It would also be the world's first alloy steel bridge. A board of 27 civil engineers condemned the plan. Meanwhile, Eads had started to build the west abutment.

Construction of the bridge at St. Louis with ribs completed and roadways begun. Source: US Library of Congress. (From: Woodward, C.M. *A History of the St. Louis Bridge* (G.I. Jones and Company, 1881).)

The Eads Bridge was dedicated on July 4, 1874. "I have haunted the river every night lately, where I could get a look at the bridge by moonlight," wrote Walt Whitman in his book, *Specimen Days in America* (1887). "It is indeed a structure of perfection and beauty unsurpassable."

Next, Eads tackled the Mississippi River itself. Below New Orleans, a periodic buildup of sand and silt blocked passage to and from the Gulf of Mexico. Often, 50 or more ships moored near New Orleans with idle crewmen, waiting for the channel to clear. Eads declared that he could use the river to remove silt from the delta. Fending off the Army Corps of Engineers who wanted to dredge a canal, Eads built jetties to restrict water flow and increase the speed of the river. The rushing water dug a channel and flushed silt into the Gulf. By 1879, the river had expanded the South Pass Channel to 300 feet wide and 30 feet deep.

In his final years, Eads promoted another ambitious project: a ship railway to link the Atlantic and Pacific oceans. Teams of locomotives on multiple parallel tracks would pull huge flat cars carrying fully-loaded vessels across Mexico's Isthmus of Tehuantepec. Eads did not have the chance to realize his railway. At the age of 66, Eads in died Nassau, Bahamas, while seeking financial support for the project.

The Bridge over the Golden Gate Strait

Sometime during the early part of the war between America and Mexico (1846-1848), Captain John C. Frémont, explorer and US Army topographical engineer, took the time to admire the three mile long and one mile wide channel that connects San Francisco Bay and the Pacific Ocean. The channel reminded him of Chrysoceras, or Golden Horn, a waterway in ancient Byzantium. So, he named California's channel Chrysopylæ, the Golden Gate. Although aesthetically pleasing, riptides and strong winds ensure that the Golden Gate Strait is not idyllic. It would take about 80 years before serious thought turned to the idea of building a bridge over the turbulent strait.

In the early 20th century, Michael M. O'Shaughnessy, San Francisco's city engineer, saw a need for such a bridge. The rapidly growing population relied upon auto transportation, serviced by the Bay's two inefficient ferry systems. O'Shaughnessy discussed the prospect of a bridge with engineers over the years. Those who thought that it would be possible typically estimated a cost of about $250 million. In 1917, O'Shaughnessy raised the subject with 47-year-old bridge builder Joseph Baermann Strauss. The two men had collaborated on the city's Fourth Street Bridge. Strauss thought that he could build a Golden Gate bridge for under $25 million.

A survey explored the best way to bridge the narrowest part of the channel, a distance of 5,357 feet above water more than 300 feet deep. Ocean storms, heavy Pacific swells, fog and high winds would place heavy demands on the bridge's design. Nature did not present the only challenge. The US War Department owned land on opposite points of the channel: the Presidio to the south and Fort Baker in Marin County to the north. A bridge might interfere with military security. The Army and Navy voiced concern that an enemy could bomb a bridge and block the Bay. The Sierra Club protested that a bridge would inflict environmental damage.

"The portion of the strait between the light house on the north and the fort on the south, is termed 'The Golden Gate,' or 'Chrysopylæ.'" Source: Shepp, James W. and Daniel B. Shepp. *Shepp's Photographs of the World* (Globe Bible Publishing Company, 1891).

Many also argued that a bridge would disrupt the beauty of the Golden Gate Strait. In her July 1924 *Harper's Magazine* article, Katharine F. Gerould hoped that the bridge project would never be realized. "[I]n the interest of your own uniqueness, dear San Francisco," she pleaded, "do not bridge the Golden Gate. Leave that kind of gesture to Los Angeles – which, if it had a Golden Gate, would most certainly bridge it, and sink oil wells into bay and ocean on either side of the bridge."

Despite objections, Strauss forged ahead, proposing a bridge that would cost $17 million and would combine the rigidity of a cantilever bridge with the light weight of a suspension bridge. The combination, Strauss believed, would fashion a bridge securely anchored to resist waves and powerful tides with a superstructure sufficiently flexible to endure strong winds. Over the next several years, increasing skepticism greeted Strauss' hybrid bridge. In 1926, the Joint Council of Engineering Societies of San Francisco requested a series of new engineering studies, which concluded that Strauss' original design had serious flaws. Others applied for the job of chief engineer.

To redesign the bridge, Strauss collaborated with structural engineer Charles Ellis, a professor at the University of Illinois and vice-president of Strauss' Chicago firm. Meanwhile, Strauss lobbied and politicked, securing the job of the Bridge District's official engineer even though he had yet to build a bridge as large as the one required to cross the Golden Gate Strait. Strauss added another member to the team: Leon S. Moisseiff, designer of New York's Manhattan Bridge. Backed by Moisseiff's calculations, Strauss recommended a suspension bridge with two flanking towers embedded in concrete piers. The bridge would have two wire cables that spanned over the towers and anchored into rock at both sides of the strait. A deck would extend between the towers, connected by suspender cables to the two main cables. The cost estimate headed toward $27 million.

Support grew for what would be the longest single-span suspension bridge in the world. During the summer of 1929, officials requested funding from the Hoover administration. The stock market crashed in October while the government still mulled over the idea. A failing economy decided the matter: San Francisco would have to raise its own funds for the bridge.

In 1930, Strauss hired John Eberson as a consulting architect. Eberson, who had created lavish Art Deco interiors in theaters, introduced Art Deco motifs to the bridge's design. The towers would have a stepped-off form of indenting segments that decreased in size with increasing distance from the roadway. In this fashion, the towers would diminish as they climbed toward the sky, a reversal of traditional bridge design. The stepped motif had been used in New York's Chrysler Building, but not in a bridge.

Although Charles Ellis had performed much of the technical and theoretical work on the bridge, Strauss told him to take an indefinite vacation without pay in late 1931. Strauss replaced Ellis with engineer Clifford Paine, who worked with architect Irving F. Morrow. Together, they altered the design to one favoring lightweight towers and curved suspension cables, and that incorporated simplified geometric patterns, aerodynamic motifs, and other hallmarks of the Art Deco movement.

Strauss lobbied again; this time, for funds to pay for the bridge. San Francisco voters approved a $35 million bridge bond, but a series of lawsuits stalled momentum for construction. The last major legal hurdle occurred in November 1931 when representatives of a San Francisco-based property development firm, the Garland Company, and the Del Norte Company, a north-state lumber organization, filed for an injunction to prohibit the Bridge District from selling its bonds. Bridge supporters claimed – with justification – that the two plaintiffs acted for the Southern Pacific, which owned 51 percent of the Golden Gate Ferries stock. In July 1932, a federal judge refused to grant the injunction. Now, bridge builders had to sell the bonds to raise money for the project. In the fall of 1932, Strauss and the directors turned to A.P. Giannini, the founder of the Bank of America. Giannini agreed that his bank would buy $6 million of the bonds, convincing others to invest in the bridge.

Construction began on January 5, 1933. The bridge would span 4,200 feet, and its towers would rise 746 feet above the water. John Van Der Zee notes in *The Gate* (2000) that Strauss' company "was charged with overseeing the building of the largest bridge of its kind ever attempted, with the tallest steel towers ever erected, the longest, largest cables ever spun, and the most enormous concrete anchorages ever poured." Yet Strauss had never supervised the erection of a suspension bridge.

As if he didn't have a sufficient number of challenges, Strauss set another: a safety record. At the time, contractors typically skimped on safety precautions to cut costs. Bucking tradition, the Golden Gate Bridge project included strict safety protocols. For more than three years, construction did not cause a single death. One of the safety innovations was the use of four enormous, manila rope nets hung beneath the bridge. Builders hung two nets from each tower and extended the nets as construction progressed toward the strait and the shores. Men saved by the nets considered themselves members of the Halfway-to-Hell Club. Despite unusual precautions, a man died in October 1936 when a crane collapsed. In February 1937, a manned 60 foot platform tore from its frame and fell thought the net. Ten men died. The company repaired the net and continued work.

A color for the bridge inflamed long-standing debates. Gray was suggested, but dropped; a gray bridge would practically vanish in a fog. The

US Air Force urged orange and white stripes so that the bridge would not pose a hazard for flying. The Navy picked white and black stripes. While the debate continued, the construction crew painted the bridge with a primer coat of pure red lead. The red color responded to changes in light and contrasted with the blues and grays of water and sky. Irving Morrow recommended orange-red for the towers and deeper shades for the approaches and cables. Orange vermilion, or International Orange, became the official color of the Golden Gate Bridge.

The bridge opened two months ahead of schedule. The city kicked off the Golden Gate Bridge Fiesta on Pedestrian Day, May 27, when an estimated 200,000 people crossed the bridge. The week-long celebration continued with parades, contests, yacht races, fireworks and fashion shows.

A painting by Irving Sinclair on the cover of the official souvenir program of the Golden Gate Bridge Fiesta.

In December 1951, nature tested the Golden Gate Bridge: A gale blew through the strait with 70 mile per hour winds. The deck of the closed bridge oscillated five feet vertically and 24 feet sideways. After the storm, inspection of the bridge uncovered only minor damage. When asked how long the bridge would last, Strauss had replied "forever."

VI. Military and Law Enforcement

Greek Fire

In the 13th century, Western European crusaders encountered a terrifying new weapon, a flaming liquid that adhered to whatever it struck and burned with fire almost impossible to extinguish. They named it *le feu grégeois*, Greek fire.

Byzantine and Arab chroniclers reported the first use of the incendiary agent in the 7th century when the Byzantines of Constantinople defended their seaport against a Muslim fleet. During the siege, Byzantines used metal tubes to propel a flammable liquid, which ignited either upon contact with water or by a fuse that burned at the tube's mouth. The oil-based substance burned on water, inspiring the name "sea-fire." For centuries, the strategic advantage of this weapon tipped the balance of power as Constantinople repelled attacks from invading forces.

Although the Byzantines held the incendiary liquid's formula a closely guarded state secret, others learned about it or invented their own versions for use on land and sea. During the 9th century, Muslim armies included naphtha troops, who catapulted barrels of liquid fire. The men in these *naffatun* units wore fireproof uniforms woven with *hajar al-fatila*, or asbestos, a fibrous rock impervious to flame. Indian and Chinese militaries also used weapons similar to Greek fire.

Apart from the devastating physical damage that it inflicted, Greek fire effectively demoralized the enemy. Near the island of Rhodes in 1103, for instance, Pisans encountered Roman ships with prows bearing the head of a beast made of brass or iron. The sailors pumped incendiary liquid through a tube that ended in the beast's open mouth where it ignited before striking enemy ships. "So that it seemed," wrote Anna Comnena in her *Alexiad* (circa 1148), "as if the lions and the other similar monsters were vomiting the fire."

At the siege of Mansura during the Seventh Crusade, Jean de Joinville witnessed Greek fire hurled from the sling of a war engine. "It looked like a dragon flying through the air," he wrote in his *Memoirs* (circa 1309). "This was the fashion of the Greek fire: it came on as broad in front as a vinegar cask, and the tail of fire that trailed behind it was as big as a great spear; and it made such a noise as it came, that it sounded like the thunder of heaven."

The origin of Greek fire remains unclear. According to one legend, an angel whispered the incendiary liquid's formula to Constantine the Great. Byzantine historian Theophanes credits the Syrian Kallinikos of Heliopolis for the creation of this new weapon. However, Constantinople chemists, who had access to knowledge from the Alexandrian chemical school, might have formulated the flammable liquid. Historians speculate that the Byzantines could have produced sea-fire by distilling petroleum into a product similar to gasoline, which they thickened to make a primitive form of napalm.

More than a new type of incendiary, Greek fire was a weapon system consisting of a flammable liquid and a delivery apparatus of cauldrons, siphons, tubes and pumps. As early as 513 AD, the Byzantines had used small siphons or syringes to squirt petroleum incendiaries. Yet it was during Kallinikos' time that the Byzantines developed a new technology for propelling large quantities of liquid fire from swiveling nozzles.

Turning a weapon like Greek fire against people eventually ran counter to notions about civilized combat. Upon penalty of excommunication, the Second Lateran Council of 1139 condemned and prohibited "that most wicked, devastating, horrible, and malicious work of incendiaries." Technological progress also played a role in eliminating Greek fire from arsenals as gunpowder-fueled artillery eclipsed incendiary weapons.

The Colt Revolver

In 1830, 16-year-old Samuel Colt served as an apprentice sailor aboard the India-bound merchant ship, *Corvo*. It was here that Colt invented the first practical single-barreled pistol with revolving cartridge chambers, a firearm that historians claim altered the course of history.

According to legend, Colt's unsuccessful attempts to shoot porpoises and whales off the Cape of Good Hope frustrated him. He found inspiration in the ship's capstan – a revolving drum used to haul the anchor – and whittled a model of a new type of gun. The firearm would have a revolving cylinder containing five or more bullets, allowing a shooter to fire a gun without reloading, a significant improvement over flintlock pistols.

About 20 years earlier, Elisha Collier of Boston had designed a flintlock with a rotating chambered breech. Collier's revolver required the user to rotate the chamber by hand. Colt invented a single-action firearm, in which the action of pulling back the hammer rotated the cylinder and lined up a new charge in the barrel.

When Colt returned home, he showed the gun model to his parents. They urged him to enlist for a 30-month whaling trip. Instead, Colt built a factory in Paterson, New Jersey in 1836. The Patent Arms Manufacturing Company designed and made the Pocket (.28 caliber), Belt (.31 caliber) and Holster (.36 caliber) five-shot revolvers, as well as two types of rifles. Before using the weapons, gunpowder and lead balls had to be loaded into the front chambers of a revolving cylinder. Separate primer was placed on the outside of the cylinder.

The US government bought Colt's firearms to supply the army engaged in Florida's Second Seminole War. However, the newly independent Republic of Texas became the company's largest customer.

Fighting a two-front war with Mexico and the Comanche Nation, Texas ordered 180 Holster model revolvers for its navy in 1839. These served in engagements against Mexico over the next four years. Many Colts were re-issued to the Texas Rangers, and earned a reputation as the perfect firearm for combat on horseback.

Outside of Texas, Colt's weapons did not prove popular. In 1842, Colt closed the company and began bankruptcy proceedings. Meanwhile, word spread about Colt's pistols. The Texas Rangers credited their Colt firearms with successful engagements against Indian forces.

When the Mexican War broke out in 1846, Colt contacted a former Rangers captain, Samuel Walker, who had been appointed a captain of the US Army's Mounted Riflemen. The two men redesigned the five-shot .36 caliber gun into a .44 caliber six-shooter. Weighing over four pounds, the Walker Colt, or Whitneyville Walker, had a nine-inch barrel that would fire both lead balls and new conical-shaped bullets.

In December 1846, Walker toted the massive firearm into the office of President Polk. The "hand cannon" made an impression; the government gave Colt an order for 1,000 revolving pistols to be delivered in three months. Now, Colt faced a new problem: He had no guns and no factory.

Colt persuaded Edwin Wesson, Eliphalet Remington and other gunsmiths to produce parts for his revolvers. He then convinced Eli Whitney Jr. to assemble parts at the Whitneyville arms factory.

In his book, *Journey Through Texas* (1857), Frederick Law Olmstead commented on the seemingly ubiquitous Colts he saw in 1853. "There are probably in Texas about as many revolvers as male adults," he wrote, "and I doubt if there are one hundred in the state of any other make."

Flush with new support for his weapons, Colt founded a manufacturing plant in Hartford, Connecticut. Colt's Patent Fire Arms Manufacturing Company opened in 1855. Colt built the world's largest arms factory and equipped it with the most modern metalworking machinery. The company earned an international reputation for extraordinary design, quality, uniformity and workmanship.

The American Civil War decreased Colt's customer base. At the start of the war, he stopped shipments to the South. The six-shot, .44 caliber Colt Model 1860 Army revolver became the Union soldiers' principal handgun.

Colt did not live to see the end of the war; he died in January 1862. He had produced more than 450,000 weapons and such memorable slogans as "God created men equal – Colonel Colt made them equal."

After the Civil War, Colt's company manufactured a breech-loaded revolver that used metallic cartridges containing a bullet, gunpowder and primer. The Single Action Army Model 1873, or The Peacemaker, became the customary sidearm of the military, the Texas Rangers and cowboys. In time, it became known as "the gun that won the West."

Alfred Nobel and Dynamite

On October 21, 1833, Alfred Bernhard Nobel was born in Stockholm, Sweden. His father, Immanuel, was an engineer and inventor. As a builder of bridges and buildings, Immanuel had an interest in new techniques for demolishing rocks that hindered progress, an interest that Alfred would share one day.

In 1837, a fire destroyed Immanuel's factory and forced him into bankruptcy. He fled Sweden, moving first to Turku, Finland and then to St. Petersburg, Russia. In St. Petersburg, he started a workshop to produce iron components for steam engines and industrial machines, as well as land mines, cannonballs and mortars for the Russian army. By 1842, Immanuel's business flourished sufficiently to enable him to bring his wife Andriette and their sons to the city. Under the tutelage of private teachers, the Nobel sons studied mathematics, physics, chemistry, literature and philosophy. Alfred excelled at foreign languages, and by the age of 17, spoke fluent Russian, English, German and French. He also developed a passion for writing. To dissuade Alfred's ideas about becoming a writer, Immanuel sent his son abroad for two years so that he could receive additional training in chemical engineering.

In the Paris laboratory of Professor T.J. Pelouze, Nobel met Ascanio Sobrero, an Italian chemist who had recently invented nitroglycerine, an explosive more powerful than gunpowder. At the time, nitroglycerine seemed too hazardous for any practical application. Many years later, Hiram Stevens Maxim, the inventor of the machine gun, described the dangers of nitroglycerin. "It was said of it that if you wished it to explode it was impossible to make it do so," Maxim wrote in *The North American Review* (February 1899). "If you handled it with great care and did not wish it to explode it was almost sure to go off; sometimes it could be set on fire, and it would burn very much like a slow fuse, while again the least jar would cause the most frightful detonation."

In 1852, Nobel returned to Russia to help the family business, which boomed during the Crimean War. The factory rapidly expanded to supply weapons and explosives to the military. Immanuel even convinced the Tsar that his little-tested invention of "naval mines" – submerged wooden casks of gunpowder – could protect the city. Immanuel was proved correct. Peppered throughout the Gulf of Finland, the mines deterred the British Royal Navy from sailing into firing range of St. Petersburg.

Immanuel lost his best customer when the war ended. Once again bankrupt, Immanuel, Andriette, and their youngest son Emil returned to Stockholm. The three older sons stayed to restructure the failed company. In a makeshift laboratory, Alfred invented three scientific devices, which brought his first patents but little income. While the brothers searched for a new moneymaker, one of Alfred's old tutors, Nikolai Zinin, reminded him about the potential of the unstable nitroglycerin. Nobel became interested in the possible use of the explosive in construction.

Around 1860, Alfred began his experiments with nitroglycerin, which he continued after returning to Stockholm in 1863. Alfred and a small group used a shed to make nitroglycerin. In September 1864, the shed exploded. Five people died, including Alfred's brother, Emil. The local authorities banned further nitroglycerin experiments within the Stockholm city limits. Determined to continue, Nobel moved nitroglycerin production to a covered barge anchored near the shore of Lake Mälaren, outside city limits.

In addition to inventing a simplified method for manufacturing nitroglycerin, Nobel was the first to devise a way to detonate nitroglycerin in a practical manner. Nobel's blasting cap would establish the basis for the modern explosive materials technology. Without significant modification, the device was used into the 1920s.

Nobel opened nitroglycerin manufacturing facilities throughout Europe and in the United States. It was a tricky business; nitroglycerin remained unpredictable. The citizens of New York City were introduced to the volatile substance one Sunday morning in 1865. A guest at the Wyoming Hotel removed a box from under the counter of the hotel's office to rest his foot while he polished his boots. "Noticing a reddish vapor emanating therefrom, he drew the attention of the hotel clerk to it, who, taking the box in his hands, made his way to the front door and threw it into the gutter, whereupon the explosion instantly followed," reported *The Manufacturer and Builder* (January 1884). "The windows of every house within one hundred yards of the entrance to the hotel were shattered, pedestrians were thrown down, and the pavement broken up." An investigation revealed that the box had been left as security by a German guest, who said that the box contained Glonoin oil – nitroglycerin – a new material for blasting.

In 1866, an explosion destroyed the Alfred Nobel & Co. factory in Krümmel near Hamburg. As he supervised the clearing of debris, Nobel considered ways to make nitroglycerine safer to handle. On the German moorlands, he found *kieselguhr,* a diatomaceous earth, which could absorb liquid nitroglycerin and ease handling. The combination of nitroglycerin and *kieselguhr* formed a doughy substance. Not only could it be shaped, but the substance could be jolted without triggering an explosion.

Nobel secured patents on the doughy explosive that he called dynamite. To detonate rods made of dynamite, Nobel invented a blasting cap that could be ignited by lighting a fuse. The construction industry readily adopted the new explosive. Along with the newly invented pneumatic drill and other innovations, dynamite significantly reduced the costs for many types of construction activities.

In time, Nobel obtained 355 patents as he devised inventions in such diverse fields as explosives, aerial photography, synthetic rubber, artificial silk, and others. Vigorously marketing his inventions throughout Europe, Nobel traveled relentlessly and spent much time in small train compartments that he called "my rolling prisons." By the early 1870s, he had made himself one of Europe's wealthiest men.

Nobel moved to Paris in 1873. An Austrian, Countess Bertha Kinsky, answered Nobel's newspaper advertisement for a "lady of mature age, versed in languages, as secretary and supervisor of household." The countess turned out to be the most qualified candidate. She worked with Nobel for about one week before returning home to marry Count Arthur von Suttner. Yet Nobel and Bertha von Suttner remained friends and corresponded for years.

While he lived in France, Nobel invented *ballistite* – smokeless gunpowder. Since the French government had allowed Nobel to use its firing ranges for his experiments, Nobel offered the government rights in the invention. The French declined; they believed that they had the same type of gunpowder, one invented by Paul-Eugène-Marie Vieille. When Nobel signed a contract with the Italian government for the delivery of three hundred tons of *ballistite*, the French gunpowder monopoly claimed that Nobel had improperly exploited the government's firing ranges. Amidst the press' accusations of treason and espionage, Nobel left the country.

Although he moved to San Remo, Italy, Nobel planned to settle in Karlskoga, Sweden. Before he could do so, however, he died on December 10, 1896.

Nobel left a bombshell in his will. In January 1897, the executors of his will, engineers Ragnar Sohlman and Rudolf Lilljequist, learned that their former employer had left the bulk of his estate to a fund. The interest would be divided into five equal parts and awarded annually to those whose work had been of greatest benefit to mankind. Four prizes would be awarded to people in the fields of literature, physics, chemistry, and physiology or medicine. The fifth portion would be awarded to "the person who shall have done the most or the best work for fraternity between the nations, for the abolition or reduction of standing armies and for the holding and promotion of peace congresses."

That Nobel would create a Peace Prize may seem incongruent. During the last ten years of his life, Nobel developed rockets, cannons, and other types of weapons technology. His inventiveness significantly advanced the escalating deadliness of military weapons. "For many hundreds of years common black powder was the only explosive used in warfare," Maxim wrote. "It was not until after Nobel invented a process for the manufacture of nitro-glycerine, that engineers began to speculate upon the possibilities of using something stronger than common black powder for charging shells thrown from large guns." Smokeless gunpowder, which enabled rapid fire machine guns, and dynamite-charged shells also increased war's casualties.

Bertha von Suttner, who emerged as a prominent figure in the peace movement, offered a clue to the apparent contradiction. In *Memoirs of Bertha von Suttner* (1910), she described a visit with Nobel in Zurich following a peace conference held in the fall of 1892. Nobel casually remarked about the profitability of the silk industry. "Perhaps dynamite factories are even more profitable than silk mills," she said. "And less innocent."

"Perhaps my factories," Nobel replied, "will put an end to war even sooner than your Congresses; on the day when two army corps may mutually annihilate each other in a second, probably all civilized nations will recoil with horror and disband their troops."

Nobel's sentiment was ahead of its time.

Smokeless Powder

Whether at land or sea, early 19th century military commanders contended with weapon fire's obscuring smoke. Gunpowder's combustion products – half gaseous and half solid – caused the problem. The solid products created smoke, fouled the inside of weapon barrels, and reduced the force of the explosion. Chemists searched for a smokeless type of gunpowder, one that burned virtually all of its materials to hot gas.

In 1846, Austrian Christian Friedrich Schöebein stabilized nitrocellulose, or guncotton, by soaking cotton waste in nitric acid. The military found Schöebein's smokeless powder too unpredictable to use.

French chemist Paul-Eugène-Marie Vieille plasticized guncotton by mixing nitrocellulose with ether and alcohol. He introduced his dependable smokeless powder, *Poudre B*, in 1886. Other inventors rushed to create their versions of smokeless powder.

In 1888, Sweden's Alfred Nobel used nitroglycerin to formulate *Ballistite*. Around the same time, US-born British inventor Hiram Stevens Maxim developed a cotton-based smokeless powder that could be extruded in cords, a type of powder later dubbed Cordite. British chemists Frederick Abel and James Dewar developed their own version of Cordite.

By 1887, the French Army had adapted a rifle and suitable cartridges for smokeless powder. The new weapon produced a higher muzzle velocity and had an improved accuracy of long range fire. The more powerful smokeless powder required smaller bullets. Now, soldiers could carry more ammunition and shoot without smoke signaling their position.

Smokeless powder ignited other developments in weapons. Since it did not produce corrosive fouling of black powder, smokeless powder enabled the development of autoloading firearms with numerous moving parts, and rapid-fire deck guns on navy ships. Smokeless powder produced a uniform gas, which could be used as a source of energy for operating a weapon. In 1884, Maxim developed such a weapon, the first true machine gun.

M'DONOUGH'S VICTY ON LAKE CHAMPLAIN.

Amidst obscuring smoke, an American warship floats between two British ships during the battle on Lake Champlain (Battle of Plattsburgh) in the War of 1812 (lithograph circa 1846). Source: US Library of Congress.

Pete Ellis: Military Strategy

More than two decades before the Second World War, Lieutenant Colonel Earl Hancock "Pete" Ellis predicted that the US and Japan would engage in an amphibious battle for the Pacific. Ellis drafted a detailed plan of attack and was researching further details when he died in 1923 on the Caroline Islands, controlled by the Japanese government. Just how Ellis died continues to incite controversy.

Ellis was born on December 19, 1880 at Iuka, Kansas. At the age of 19, he joined the US Marine Corps and advanced to second lieutenant a year later. In 1911, Ellis took a course at the Naval War College (Newport, Rhode Island). Upon request, he stayed as a lecturer for a year. While at the college, Ellis wrote papers about the strategic importance of naval bases. Before the US entered the First World War, Captain Ellis performed a special terrain study and intelligence service in the West Indies and at the Naval Station on the island of Guam. While on Guam, Ellis and a small group of men showed that artillery could be landed from boats.

In February 1918, Ellis, now a major, was assigned to the Office of Naval Operations in Washington, D.C. During the summer, he was transferred to the staff of General John A. Lejeune and served as an observer of operations in France. Soon, Lieutenant Colonel Ellis participated in the planning and execution of offensive actions. His role in military operations earned Ellis France's Croix de Guerre and the Légion d'Honneur. US President Woodrow Wilson awarded him the Navy Cross.

After the war, Ellis became head of the intelligence section in the Marine Corps' new Division of Operations and Training. He soon turned his attention to Japan. The Treaty of Versailles awarded Japan the Carolines, Marshalls, and Northern Marianas, which the Japanese Imperial Navy had wrested from Germany's control in 1914. Japan promised not to fortify the islands. Yet Ellis foresaw a war with Japan, and he knew that the islands would provide Japan with bases that could be used to launch attacks on the Philippines and other American possessions in the area. In 1921, he wrote *Operations Plan 712-H: Advanced Base Operations in Micronesia*. The document detailed a strategy for the use of Marines as amphibious shock troops to quickly implement surprise offensive actions to capture the island naval bases. The occupied bases would open a corridor for the US Navy to advance against Japan. His tactics departed from tradition. In the past, military planners had delegated the role of the Marine Corps as defenders of advanced bases to support the Navy, as if those bases would be readily available for occupation.

To prepare for the possibility of war, Ellis needed more information about islands controlled by Japan, including the possible existence of fortifications and a picture of the terrain. During the Roaring Twenties, no satellite circled the globe to provide such data. Intelligence about the Micronesian Islands had to be performed by someone on the ground. Ellis traveled to the islands posing as a travel agent buying dried coconut for the John A. Hughes Trading Company of New York City. Before leaving the US in August 1921, Ellis obtained the latest navy code from the Office of Naval Intelligence and pasted it in the back of a tattered copy of *Bentley's Business Code*. In May 1923, Japanese authorities reported that Ellis had died on the Micronesian island of Korror. The press speculated that Japanese secret police had executed Ellis. To this day, Ellis' death fuels conjecture.

"What actually happened is not mysterious," wrote Dirk Anthony Ballendorf in his article, "Earl Hancock Ellis" (*Micronesian Journal of the Humanities and Social Sciences*, December 2002). Ballendorf examined official records and obtained statements from those who had known Ellis, or had met the man during his trip across the islands. Ballendorf learned that Ellis had been in and out of hospitals on his journey, treated for delirium tremens. Although his health had deteriorated, Ellis had refused medical advice to stop drinking. He had become violently ill days before he died.

Twenty years after his death, the US military implemented Ellis' strategy for capturing bases in Micronesia. The actual campaign deviated from the original plan only to accommodate technological advances.

"Although he was no means alone in recognizing that the balance of power in the Pacific had shifted with Japan's acquisition of Micronesia," Ballendorf said, "his unique contribution was that he knew what the Marine Corps should do about the threat, and he acted on that belief."

Radar

US Navy Lieutenant Commanders Samuel M. Tucker and F.R. Furth coined the term "radar" for RAdio Detection And Ranging. Yet researchers in many countries developed the technology – almost simultaneously. Radar uses reflected high frequency radio impulses to detect and locate objects. Early forms of the technology focused on detection.

Inspired by the death of a friend who died on a ship collision, German inventor Christian Hülsmeyer created an anti-collision device. Hülsmeyer's *Telemobiloskøp* worked like a searchlight that used radio waves instead of visible light. On May 18, 1904, he demonstrated that his invention could emit radio waves at, and receive reflected radio waves from, a ship approaching the Hohenzollern Bridge in Cologne. Although the press and the public applauded the *Telemobiloskøp*, industry and the Navy greeted Hülsmeyer's invention with disinterest.

Guglielmo Marconi revitalized the anti-collision idea in a 1922 speech. He explained that radio waves could be reflected by metallic objects located miles away. Perhaps a ship with suitable equipment, Marconi suggested, could detect the presence and location of other vessels.

During the same year, Albert Hoyt Taylor and Leo C. Young experimented with high frequency radio communications for the US Naval Research Laboratory. A wooden steamer, the *Dorchester*, happened to sail up the Potomac River and pass between their transmitter and receiver. When Taylor and Young realized that the *Dorchester*'s passage interfered with their signal, they proposed a system of receivers and transmitters that would detect enemy vessels attempting to slip into harbors at night.

As uses for radio transmissions increased during the 1930s, many noticed the phenomenon of radio signal interference. Taylor and Young discovered that aircraft interfered with radio signals, a finding that motivated the development of modern radar. While testing a microwave radiotelephone link for the Pope, Marconi observed that a moving steamroller disturbed a signal. This triggered Marconi to construct a radar prototype. In Great Britain, reports about radio interference instigated a radar program that played a crucial role in the outcome of the Second World War.

In 1934, the Director of Scientific Research for the British Air Ministry asked Robert Watson-Watt if he could develop a radio "death ray" to incapacitate an enemy aircraft pilot. Watson-Watt, superintendent of the National Physical Laboratory's Radio Department, passed the request to Arnold F. Wilkins, who calculated that current technology could not generate sufficient power for such a weapon. Wilkins knew that government post office engineers reported disturbances caused by airplanes flying near their shortwave communication receivers. He suggested that they might be able to use the phenomenon to detect aircraft.

Instead of a death ray, Watson-Watt offered the Air Ministry a Radio Direction Finding system, highlighting that Radio Research Station investigators had devised the basic technology for their studies of the ionosphere. Watson-Watt and his colleagues demonstrated the feasibility of the proposal using the British Broadcasting Corporation's shortwave radio transmitter at Daventry to detect a Royal Air Force bomber. This successful test led to the installation of a chain of radar stations along England's coasts before the outbreak of war in 1939. The Chain Home stations provided critical information that guided the relatively small number of RAF fighter pilots to the most advantageous locations in their battle against the *Luftwaffe*.

The British development of radar provides one example illustrating that the technology's time had arrived. Increasing international tensions during the 1930s spurred efforts to develop radar technology in Canada, France, Germany, Italy, Japan, the Netherlands, the United States and the USSR. Radar research continued after the Second World War. Today, radar aids geological exploration, severe weather tracking and the ticketing of speeding motorists.

Barbed Wire

The US Homestead Act of 1862 opened vast amounts of public land to citizens willing to build a home and to farm for five years. Many homesteaders of the American West found that survival required a new type of barrier, a need met by the invention of barbed wire.

When pioneer-farmers traveled west, they learned about an unwritten rule of free access to grass and water for cattle on unoccupied government lands. The range cattle industry relied upon the Law of the Open Range, because few cowmen owned their grazing land. To protect their crops from wandering livestock, the new farmers had to construct fences. This posed a challenge in the prairies and plains, which offered scarce amounts of timber and rock.

For decades, farmers had tried fences made of smooth wire and wooden posts. The wire alternative to a traditional fence proved inadequate in the West, where wire snapped in bitter cold winter, sagged in scorching summer, and provided little resistance to foraging cattle and sheep.

Beginning in the late 1860s, American farmers revived the practice of creating thorny barriers with honey locusts, thorn locusts, Osage orange, briar and other prickly plants. Yet hedge-planting had serious drawbacks: Hedges grew too slowly, took up space and harbored rodents.

A man named Rose spurred the growth of a new industry when he created an artificial thorny hedge. At the 1873 county fair in DeKalb, Illinois, Henry M. Rose displayed a wooden rail bearing short wire projections, which he designed for attachment to a fence. Inspired by Rose's innovation, Joseph Farwell Glidden fabricated coiled pieces of pointed wire with a modified coffee grinder, and threaded these coiled barbs on a strand of plain wire. Six months after visiting the fair, Glidden filed his first patent application on barbed wire.

In November 1874, Glidden received a patent for an improved barbed wire consisting of the now-familiar double twisted wires and barbs. After Glidden sold half of his patent rights to Isaac Leonard Ellwood, the two men established the Barb Fence Company in DeKalb. Their business became the heart of a multi-million dollar industry that produced fencing touted by promoters as "cheaper than dirt and stronger than steel."

Farmers of the American West initiated a lasting cultural transformation by installing barbed wire fences throughout formerly open terrain. The farmers' fences curtailed the lifestyle of nomadic Native Americans, who found themselves barred from their ancient territories. Dependent upon dwindling public lands, cattle herders initially opposed the "devil's rope." Half a century later, Cole Porter's song, "Don't Fence Me In," would reflect a common reaction to barbed wire in the Old West.

During the 1880s, many cattlemen of the Southwest changed their minds about barbed wire when they decided to possess their own ranges. Cattlemen not only fenced land that they owned, but also land they intended to lease or purchase, including public lands administered as Open Range. This fence-first practice incited violent opposition. Fence Cutter Wars erupted in New Mexico, Texas, Wyoming and other states. In Texas, the battles ended after fence cutting became a felony, marking the end of free range for Texan cowmen and the birth of the contained cattle ranching industry.

Barbed wire also played significant roles in wars between countries, including the Spanish-American War, Boer War, and Russo-Japanese War. During the First World War, soldiers strung barbed wire along trenches to impede infantry charges on the Western Front, a strategy that contributed to years of deadlock.

Today, artists incorporate barbed wire in sculptures and jewelry. It has also become a collectable. Enthusiasts have catalogued over 2,000 variations of the prickly wire.

German soldiers fix a barbed wire entanglement during the First World War.
Source: US Library of Congress, Prints & Photographs Division, LC-B2- 3376-9.

The Polygraph

Techniques for identifying deception have often relied upon physiological markers for stress, such as decreased saliva production. The hill tribes of Rajmahal, for example, required suspected liars to lick a hot iron. A truthful person would pass the test with an unburned tongue. The ancient Chinese required suspects to place rice powder in their mouths and spit it out. If a person had been lying, then the powder remained dry. During the Inquisition, subjects had to swallow a "trial slice" of bread and cheese, a difficult feat to perform with a parched mouth.

During the end of the 19th century, Cesare Lombroso, considered the father of psychological profiling, noticed that a rise in blood pressure and pulse rate accompanied the telling of a lie. The Italian criminologist constructed the hydrosphygmograph, a device that recorded a subject's pulse as a line on a revolving drum. In 1895, Lombroso claimed that he had detected lies with his machine.

Austrian psychologist Vittorio Benussi took a different approach to determine whether a person was being economical with the truth. By 1914, Benussi had concluded that changes in respiration rate signaled deceit. The ratio of inspiration to expiration, he decided, is usually greater before truth telling than before lying.

Several years later at Harvard University, William Moulton Marston began his research into uncovering deception. Like Lombroso, Marston became convinced that he could detect verbal deception with a machine that measured an increase in blood pressure. Marston's testing of a murder suspect shaped the judicial view of lie detector evidence in US courts.

In November of 1920, someone killed Dr. R.W. Brown. After several days of interrogation, James Alphonzo Frye confessed to the murder, but repudiated his admission just before trial. Frye's attorney called upon Marston, who administered his blood pressure lie test and concluded that the defendant was innocent. But the judge would not allow Marston to present these results to a jury. An appellate court agreed with the judge, and the case, *Frye v. United States*, set a standard for admissibility of scientific evidence that stood for more than half a century. Despite this defeat, Marston remained an advocate of his lie detector. His fascination with thwarting deceit can be seen in a creation penned under the pseudonym Charles Moulton: the comic book hero Wonder Woman, whose golden lasso compelled villains to speak the truth.

The ability to detect lies had been on the mind of August Vollmer, Berkeley, California's progressive Chief of Police. After reading an article by Marston in 1921, Vollmer discussed lie-detecting machines with Sergeant John A. Larson. The sergeant, who would later earn a medical degree, built a variation of the inflatable rubber tube apparatus that doctors used for blood pressure measurements. Larson called his machine the cardio-pneumo-psychogram. But it soon became known as the lie detector.

Leonard Keeler, another recruit to the Berkeley police force, redesigned Larson's detector. First, he added a respiration-checking device, a modification that reflected Benussi's strategy. In 1926 he attached a galvanometer component to measure changes in skin resistance caused by perspiration. The polygraph was born.

Today's computer-assisted polygraph also measures variations in blood pressure, heartbeat, pulse, skin resistance and breathing. In North America, law enforcement agencies primarily use the polygraph as an investigative tool, not for collecting evidence. Judges, who remain skeptical about the evidentiary value of polygraph test results, prefer to leave the detecting of lies to jurors.

Truth Serum

A truth serum is neither a serum nor a reliable means to secure the truth. Honest.

The notion that a drug could force truthful statements dates to a time when someone first realized that inebriation inspires a person to talk. Ethanol – consumed as a beverage or injected – became a type of truth drug.

In the early 20th century, Robert E. House, a Texas obstetrician, ushered in the era of the modern truth drugs with scopolamine, a substance that reduces inhibitions and makes a subject more talkative. For years, obstetricians had used the drug to help their patients during labor. Scopolamine did not lessen pain. Rather, the drug induced a half-conscious twilight sleep and blurred the memory of the pain experience.

Robert House discovered a new effect of scopolamine. In 1916, he asked his patient's husband to fetch a set of scales to weigh a baby, but the husband could not find them. The scales are in the kitchen on a nail behind the picture, advised the new mother, deep within a twilight sleep. Amazed that the woman offered reliable information while severely unconscious, House performed additional tests. He became convinced that, under the influence of scopolamine, a person lacked the ability to think or reason and could not craft a lie.

On February 13, 1922, House agreed to a request by Dallas district attorney Maury Hughes to publicly test his method. The doctor would try to extract the truth from two prisoners residing in a local jail.

W.S. Scrivenor, one of the test subjects, later described the effects of the drug. "Answers to questions slipped from my mind without any apparent desire to stop them," Scrivenor told *The Galveston Daily News*, "I felt that I couldn't formulate any imaginative trimmings to them."

Hughes became ecstatic about the results, according to *The Bee* (Danville, VA). "If we can buy truth in bottles and inject it into criminals' veins," he said, "the lie – the criminal's best defense – will be useless!"

Within a few months, a newspaper reporter coined the term "truth serum." It stuck and captured the imaginations of readers who had a great confidence in science and a fascination with new forensic science techniques. The application of science had achieved so much in recent decades. Why shouldn't a drug reveal truth?

During the 1930s, law enforcement officials embraced truth serum as a vital tool to acquire confessions from obstinate criminals in an extensive attack on organized crime. Narcoanalysis, the extraction of information with truth drugs, replaced the violent third-degree interrogation.

Scopolamine became outdated at this time; the drug had many undesirable side effects, including hallucinations and drowsiness. Truth seekers turned to barbiturates, such as sodium amytal and sodium pentothal.

The US experienced a widespread use of truth serum into the 1950s. Truth serum was not only popular in police departments. The military and intelligence agencies used truth drugs in interrogations of spies and prisoners during the Cold War and the Korean War.

Around this time, however, truth drugs began to lose their allure. Researchers reported that, despite a truth drug's influence, a significant number of subjects managed to mislead or lie to their interrogators. Scientists denied the validity of the entire truth drug concept. US judges also did not care for drug-induced testimony. They found drug-influenced statements inadmissible as evidence to establish truth.

Nevertheless, truth drugs still make headlines. Sodium amytal and sodium pentothal have been used to recover repressed memories – with decidedly mixed results. And in the wake of terrorist attacks, US officials reconsidered the application of truth drugs for interrogations.

Forensic Firearms Identification

Forensic firearm examiners determine whether a certain weapon had fired a bullet or cartridge found at a crime scene. Early efforts linked spent ammunition with a class of weapon. Following the 1862 shooting of Confederate General Stonewall Jackson, for example, investigators concluded that the General had been accidentally shot by his own side. The spherical projectile removed from the General had been fired by a smooth-bore musket, a type of weapon that the Union Army no longer used.

In 1912, Professor Victor Balthazard at the University of Paris formulated the basic principles of firearms examination. Using enlarged photographs, he compared marks created by a firearm on the surface of bullets and cartridge cases found at a crime scene with marks on ammunition that he had fired from a suspect weapon. In this way, he could connect crime scene ammunition to a particular firearm.

During the 1920s in New York, four men rediscovered Balthazard's principles and initiated modern firearms identification: Charles E. Waite, Calvin Goddard, Philip O. Gravelle, and John E. Fisher. Gravelle had substantial experience with a comparison microscope to study fine details in cloth patterns. He suggested that they might be able to use the instrument to compare fired bullets and cases.

In the signal event of firearms identification, the group bought two comparison microscopes and modified them. They added a comparison bridge, and rotatable mounts for bullets and cartridge cases. Through the eyepiece of the bridge, two pieces of spent ammunition could be examined, one on each stage of the two microscopes.

Police departments and the courts became aware of the value of "fingerprinting" bullets, especially after Goddard testified about his findings in the 1929 St. Valentine's Day Massacre. A decade later, firearms identification had become an established technique of criminal investigation.

Expanding Bullets

New smokeless gunpowder of the late 19th century increased the speed of bullets and improved a firearm's accuracy. To accommodate greater speeds, bullets had to be smaller and covered with metal to reduce stripping. British soldiers discovered that the new bullets had a reduced penetration power and produced neat, clean punctures. Their ammunition could not stop the furious charges of Pathan tribesmen on the northwest frontier of British India.

Historians credit Major General William Tweedie with devising a notorious way out: the dum-dum bullet. The British rifle and ammunition factory at Dum Dum in India manufactured a new type of bullet that had a similar weight to the standard round-nosed bullet. But, it also had a 0.5 mm opening in its jacket to expose the lead core. With its weakened metal jacket, the bullet expanded after impact to create a large, devastating wound.

The British Army issued several million pounds of dum-dum ammunition. In 1897 and 1898, soldiers used the bullets during the Chitral and Tirah expeditions on the Northwest Frontier. In 1899, the International Conference of The Hague prohibited the use in warfare of bullets that readily expand in the body.

Today, hunters and police forces use expanding bullets. While the design of expanding bullets changed since the dum-dums, their function remains the same. After penetrating the target, the bullet's covering peels back upon itself. The mushroomed bullet inflicts significant damage internally. After expending energy inside the body, the bullet may remain inside or stop soon after clearing the target. In police work, the expanding bullet offers a high likelihood of incapacitation. The ammunition also reduces the risk of shooting a bystander with a bullet that penetrates the target or that ricochets off a hard object.

Military Tanks

The idea of a propelled armored vehicle dates at least to Leonardo da Vinci's 15th-century sketches. Four hundred years later, American E.J. Pennington sketched his version of a steam-powered armored car with wheels protected by metal covers. In the early 20th century, F.R. Simms, an English engineer, built such an armored car.

The First World War focused efforts to develop a workable military tank. In the deadlock of trench warfare, a few machine guns killed thousands of soldiers as they struggled through barbed wire to cross no-man's land. Historians credit Britain's Lieutenant Colonel Ernest Dunlop Swinton for suggesting that an armored vehicle would break the impasse.

The British Navy's W.G. Wilson and William A. Tritton, managing director of an agricultural machine company, devised a vehicle with a rhomboid shape that had a tread circulating around its length. The machine's armor would enable drivers to face machine gun fire, while the shape and tread allowed it to cross trenches. The military ordered 100 of the vehicles, codenamed "water tanks."

The first British tanks weighed about 30 tons, traveled at a walking speed and bore machine guns and cannons. With 53 gallons onboard, the tanks had a range of 12 miles.

On September 15, 1916, the British unveiled 36 tanks at the Battle of the Somme. They gained 3,500 yards – the longest one-day British advance of the battle.

By the end of the war, the French introduced Renault-built, light weight tanks, the US Army developed its own Tank Corps, and the Germans rolled out their 148 ton tanks onto the battlefield. The introduction of armored vehicles rendered trench warfare obsolete.

J. Carl Mueller's poster advertising a fundraising event, showing the British tank, "Britannia," circa 1917. Source: US Library of Congress.

Synchronized Machine Gun Fire

At the start of the First World War, aircraft scouted terrain and fliers passed on information to ground forces. Aircraft pilots, sometimes accompanied by an observer, served for reconnaissance, not as fighters. Soon, fliers did more than observe; they shot at enemy pilots with rifles, shotguns and pistols. Sometimes, they threw hand grenades, heavy steel darts and even bricks. This did not prove to be a very effective way to fight.

Tractor aircraft had a propeller in front of the plane. A two-seater tractor aircraft could carry an observer who fired a hand-held machine gun to the side and behind. If he fired forward, he would shoot his own propeller. After a few months, pilots demanded a fixed machine gun that faced forward.

French flier Roland Garros asked Raymond Saulnier to modify a plane with steel wedge deflector plates that attached to propeller blades. Garros also had a fixed machine gun mounted in front of his cockpit. On April 19, 1915, a rifle shot nicked the fuel line of Garros' aircraft over Courtrai. German soldiers prevented the pilot from destroying his plane.

Anthony Fokker, the Danish aeronautical engineer who built aircraft for the German Air Service, experimented with the deflector plates. He found that the plates could not stand against German steel-jacketed bullets. Instead, Fokker designed synchronization gear that linked the engine's crankshaft to the firing of the machine gun. The mechanism enabled the machine gun to fire only when the propeller blade cleared the barrel.

In the autumn of 1915, Fokker Eindeckers armed with synchronized Spandau machine guns safely fired at Allied planes through the arc of spinning propellers. The "Fokker Scourge" marks a time when the German aircraft ruled the skies of the Western Front.

An Eindecker pilot unintentionally ended the scourge in 1916. Lost in heavy fog, he landed in France. Soon, the British and the French flew aircraft fitted with machine guns, synchronization technology and more powerful engines.

Sonar

The fact that sound travels through water has been known for thousands of years, dating at least to the time of Aristotle. Several years before the outbreak of the American Civil War, Matthew Fontaine Maury proposed a use for this phenomenon. He suggested that sound might be used to measure ocean depth.

The 1912 *Titanic* disaster spurred efforts to use sound for detecting objects in the ocean. Within two years, English meteorologist Lewis Richardson filed a patent application for an underwater echo ranging device, German physicist Alexander Behm obtained a patent for an echo sounder, and Canadian inventor Reginald A. Fessenden built an oscillator system that detected an iceberg several miles away.

In 1915, French physicist Paul Langévin collaborated with Russian émigré scientist, Constantin Chilowski, to generate high-pitched ultrasonic waves with quartz compression. This would be used in an apparatus to send and receive underwater echoes. Careful interpretation of the returning signal allowed the calculation of the distance and location of a target. They completed their invention, which would become known as active sonar, during the last year of the First World War.

A secret 1916 British effort under Canadian physicist Robert W. Boyle focused on ultrasonics and quartz oscillators for use in submarine warfare. A year later, the group successfully tested a prototype active sonar device. The system, built for the Anti-Submarine Division, acquired the designation, ASDIC.

By the end of the First World War, the United States had designed its own active, echolocation system. During the Second World War, the US system assumed the label, sonar, for *So*und, *Na*vigation and *R*anging.

During that war, the British ASDIC system proved crucial in efforts to impede attacks by German submarines on warships and vital supply ships.

Today, the uses of sonar technology extend far beyond military applications. Sonar is used to map the ocean floor, locate sunken ships, detect schools of fish, inspect the structure of dams and aid police investigations.

Military Snipers

In 18th century Great Britain, a snipe hunt was not a practical joke; it was a hunt for a snipe. With its mottled brown camouflage, its startling flushes and erratic flight, the snipe challenged flintlock-toting huntsmen. The triumphant hunter of the elusive bird would be hailed as a sniper.

The British military adopted the term in early 19th century India. Sharpshooters became known as snipers. British newspaper reporters popularized the word when they used it in descriptions of the Second Boer War. The Boer snipers' long distance marksmanship accounted for many British casualties. Officers proved to be a favorite Boer sniper target, a practice followed by today's military snipers.

In the trench warfare of 1914, the British and their allies faced a new threat. The Germans had formed *Jäger* unit soldiers armed with telescope-sighted Mauser rifles. These *scharfschütze* concealed themselves in unpredictable locations and hunted enemy soldiers with officers an especially preferred mark.

The British soon started their own sniper program, recruiting from large estates of England and Scotland. Here, they found gamekeepers who had the required skills of marksmanship and patience, as well as the ability to conceal themselves and observe.

Scots gamekeepers established a camouflage outfit that remains associated with the military sniper: the ghillie suit. To make the suit, they attached to the back of pants and a jacket strips of burlap sacking or hessian, usually dyed in brown and green colors. The suit included a hood covered in cloth strips with fine netting to conceal the face. To complete the ghillie suit's misshapen form, they attached bits of local vegetation.

The sniper has become an integral, valuable part of the military. Although deadly for the enemy, the sniper's primary mission often focuses on reconnaissance.

VII. Potpourri

Teddy Bears

"Success has many fathers," an old saying goes. The teddy bear certainly had several.

The birth of the American teddy bear dates to President Theodore Roosevelt's trip to the South in 1902. The governors of Louisiana and Mississippi invited the President to resolve a border dispute. In November, the President's Southern hosts took him on a five-day bear hunting expedition near the town of Smedes, Mississippi. For days, crafty bears eluded the hunters.

On the last day of the hunt, according to legend, guide Holt Collier decided to ensure that the President would take home a trophy. Collier and his dogs tracked down a 235-pound black bear. The guide stunned the bear with a club, strapped it to a tree, and offered the creature to the President. Roosevelt refused to shoot the defenseless animal.

Clifford K. Berryman, a *Washington Post* staff artist, captured the incident in a cartoon: an armed President with his back turned to a cowering bear. The caption, "Drawing the Line in Mississippi," referred to the controversy that provoked the President's visit and his refusal to shoot.

Berryman's original cartoon showed an exhausted, mature bear. The illustrator redrew his cartoon, transforming the large bear into a quaking cub.

Self-portrait of Clifford Berryman with his two bears, 1904. Source: US National Archives.

Newspapers across the country printed Berryman's revised cartoon. The image inspired Morris Michtom, a Russian immigrant who owned a small candy and notions store in Brooklyn. He asked Rose, his wife, to make a toy bear cub. She stuffed a piece of plush velvet into the shape of a bear and added shoe button eyes. Michtom placed the doll and a copy of the cartoon in the window of his store.

Unexpectedly, Michtom received more than a dozen offers for the bear doll. Not wishing to offend the President, Michtom sent Roosevelt the bear as a gift for the President's children. He also asked Roosevelt for permission to use his name for the toys. Roosevelt gave his consent, warning that he doubted that it would increase the sale of the dolls.

The Michtoms began production of stuffed Teddy's Bears. In 1903, when they could not keep up with demand, Michtom started the Ideal Toy Company.

Several stories explain the origin of the German teddy bear. All focus on Margarete Steiff, a toy manufacturer confined to a wheelchair by polio. Some predicted that young Margarete would depend upon others for her entire life. She proved them wrong.

Steiff learned needlework, and then taught others. She used the money to buy a sewing machine and made dresses. By 1879, she started a dressmaking business. Within four years, she supplemented her products with sewn poodles, elephants and donkeys.

According to one account, around the end of 1902, an American visited the Steiff factory. He showed Berryman's cartoon to Margarete Steiff, and suggested that her company should make toy bears. Steiff didn't pause; her company produced the bears and debuted them at the 1903 Leipzig Toy Fair.

In a different version, Margarete's nephew, Richard Steiff, watched a troupe of performing bears at the Stuttgart Zoo. The sight inspired him to design a toy bear jointed like a doll. Around 1902, he showed his aunt sketches of the bear doll, and Margarete designed her own jointed bear with glass eyes and mohair plush fur.

Richard displayed Steiff's Baer 55PB at the 1903 Leipzig Toy Fair. The toy's odd name referred to its features: 55 centimeters, plush (*plüsch*) and jointed (*beweglich*). The doll stirred little interest among European buyers. Bearing in mind the growing Teddy's doll craze back home, a buyer from a New York firm placed an order for three thousand. Within a year, the Steiff bear had become a hit in America.

Other companies joined Ideal Toy Company and Steiff in the US market. Soon, teddy bears were everywhere – in the arms of children and society ladies, starring in children's books and songs, and appearing on roller skates. The original Teddy bear, which now accompanied Roosevelt in Berryman's cartoons, entered politics. The bear served as a mascot in Roosevelt's 1904 presidential campaign.

After the First World War, new teddy bear companies developed in England, France and Australia. Glass eyes replaced buttons. Mechanical teddy bears made music, walked and danced.

Following the Second World War, the trend toward quality and sophistication reversed with the introduction of cheap, mass-produced teddy bears.

Soon, the demand for handcrafted dolls hibernated. The 1970s saw an awakening of the market for high-end dolls in the form of artist-designed bears. Limited edition artist bears maintain their popularity.

Clothes Washing Machines

"How would it simplify the burdens of the American housekeeper to have washing and ironing day expunged from her calendar!" declared Catherine E. Beecher and Harriet Beecher Stowe in their book, *American Woman's Home* (1869). They suggested that communal laundries might help housewives. "Whoever sets neighborhood laundries on foot will do much to solve the American housekeeper's hardest problem."

Somebody had set up a public laundry in 1851 during California's gold rush. The washing machines had been donkey-powered. Offloading the responsibility of clothes washing didn't offer a lasting answer to the dilemma. A clothes washing machine for the home, on the other hand, now, that could work.

By the early 19th century, many North American households had a simple hand-operated clothes washing machine. In some, a crank turned a wooden box that held water, soap and clothes. Others emulated time-honored methods of washing, such as a lever-operated rocking scrub board that scoured clothes between two washboard-like surfaces. These machines did not save much labor. Somebody had to power the gadget and haul up to 50 gallons of water.

The need for a clothes washer inspired inventors. By 1875, thousands of patents had been filed for clothes washing machines. These included devices that used a rake to drag clothes back and forth through soapy water, and a machine that beat clothes clean with a component shaped liked an upside down stool. Somebody even engineered the Locomotive, a washing machine that darted back and forth on a short rail, battering clothes and soapy water against the sides of its tub.

In the late 1800s, washers came equipped with a wringer. No longer would it be necessary to twist wet clothes by hand to remove excess water.

The turn of the century brought metal tubs and a motor. In 1907, the Hurley Machine Company of Chicago marketed Thor, the first electric washing machine. The apparatus had an electric motor that rotated a galvanized tub. The use of a motorized washing machine could be a shocking experience; the electric motors lacked a waterproof covering. A wet motor could spark a fire, a short circuit, or even an electrocution.

Electric washing machines still required a lot of work. Housewives had to make sure that tangled clothes did not burn out the motor. They also had to fill and empty the water, and squeeze wet clothes through a wringer.

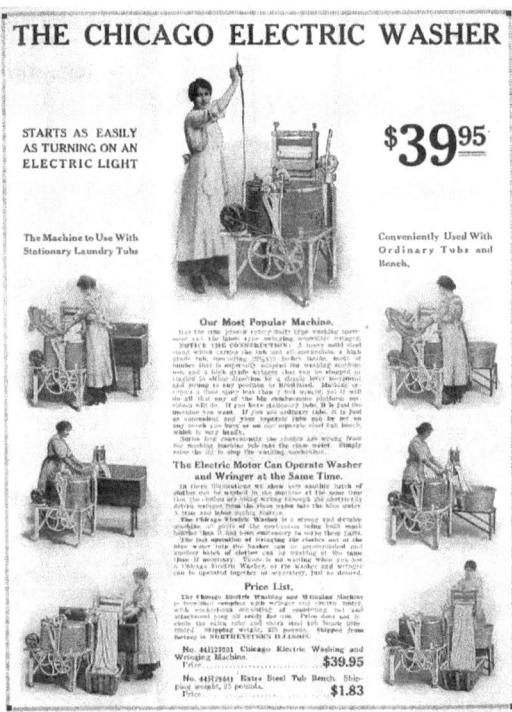

A clothes washing machine that probably did not save much time. Source: Sears, Roebuck and Company catalog, 1917.

A market for electric clothes washing machines slowly developed in America. Many houses lacked electricity. Hand- or gasoline-powered wringer-washers offered a practical solution. This changed by the late 1920s when about 84 percent of households had electricity.

New innovations of this time included a pump that filled and drained the washer's tub. The introduction of electric water heaters for the home ensured a supply of hot water. With the increased popularity of washing machines, companies protected their customers by encasing motors in waterproof shields.

During the mid-1930s, John W. Chamberlain of the Bendix Corporation invented a machine that could wash, rinse, and pull out water from wet clothes. The apparatus used a spin cycle that wrung water from clothes with centrifugal force.

Washing machine motors of the time operated at a speed dictated by the washing cycle, a speed much slower than that required for efficient drying. The 1950s and 1960s brought the two-speed motor that ramped up the spin cycle, and a timer. The timer developed from a simple signaling device into a component that controlled the duration of cycles.

Dry Cleaning

"Dry cleaning is so called because water is not used as a cleaning agent," explained L. Ray Balderston in *Laundering* (1914). "The principle involved is to use some material which is a perfect solvent for grease, and as the grease is dissolved the dirt is naturally set free." Even in 1914, dry cleaning was not cutting edge technology; it had been practiced for thousands of years.

Clay tablets from the ancient Greek city of Mycenae list occupations that include dry cleaner. The tablets may date to 1600 BC. Eschewing water and soap, the earliest dry cleaners removed stains with ammonia or lye. They also used clays or soils, known as fuller's earth, to absorb grease from clothing too delicate for water laundering.

In the 1840s, the company of Jolly-Belin became Paris' first dry cleaning business. According to legend, Jean-Baptiste Jolly had been inspired to open the firm after he saw that camphene from an overturned lantern eradicated tablecloth grease spots. Soon, dry cleaning businesses opened throughout Europe and North America, taking advantage of a chemical industry that supplied benzene, camphene, gasoline, and other organic solvents.

In 1869, the Scottish company, Pullar's Dyeworks, introduced machinery to perform dry cleaning. The increased efficiency boosted the popularity of dry cleaning, which unlike cleaning with hot water, did not weaken fabrics or fade water-soluble dyes. Not everyone was convinced about the process. The March 22, 1884 edition of *Household Words*, a publication edited by Charles Dickens, carried an article on spring cleaning tips. After distinguishing dry cleaning and thorough cleaning – "what our grandmothers called scouring, plunging the article into a tub of soap-and-water, washing it well, and rinsing in clean water" – the writer cautioned readers that the results of dry cleaning depended upon the quality of cloth. "If the materials have originally been first-class, dry cleaning is successful, but if cheap and fourth rate, are proportionally nasty."

Professor W.F. Sudro described state-of-the-art dry cleaning, circa January 1918, in *The Bulletin of Pharmacy*. Using a tumbler washing machine, the cleaner first soaked garments in a solution of gasoline and soap. After replacing the dirty gasoline mixture with fresh gasoline, the cleaner agitated the garments, and then transferred them to an extractor that removed excess gasoline using centrifugal force. Clothes were dried by steam coils or in a dry-room tumbler, a machine similar to modern clothes dryers. Sudro pointed out a hazard in a section entitled "When the Machines Blow Up." Clothes accumulated static electricity, he warned, as they rub against each other in the washer. Touching the charged, gasoline-soaked garments could spark a serious fire or explosion. The dependency upon flammable solvents, such as gasoline, ether, and chloroform, presented a serious drawback to dry cleaning of the time.

In 1920, German dry cleaners began to use trichloroethylene, a nonflammable solvent. The cleaner had its own problem: The powerful chemical damaged synthetic fibers. The industry found a solution in a very old chemical. A century earlier, British scientist Michael Faraday had synthesized tetrachloroethylene, or perchloroethylene (perc). During the 1930s, the dry cleaning industry switched to perc, a nonflammable solvent that was gentler on synthetic fabrics. Petroleum shortages of the Second World War also increased the popularity of perc over petroleum-based cleaners.

As perc, widely considered a safe solvent, became the industry standard, dry cleaning businesses opened in commercial districts and residential areas. During the 1970s, however, studies percolated about the harmful effects of the solvent on the environment and human health. While the vast majority of dry cleaners still use perc, a number of states in the United States have passed plans to phase out the solvent.

Parking Meter

Oklahoma City merchants faced a dilemma in 1933. Employees of downtown businesses arrived early and parked for long periods of time, blocking spaces that shoppers could have used. Police officers regulated parking areas by chalking the tires of parked cars and hunting for marked tires a few hours later. This arrangement tied up the police. At the same time, a driver could easily defeat the system by moving the car over the chalk mark or by erasing the mark.

The Chamber of Commerce Traffic Committee tasked Carl MaGee, a newspaper editor, to come up with a solution. He succeeded.

In December 1932, MaGee filed a patent application for a coin-operated parking meter that looked like a bread loaf balanced on the top of a pole. MaGee followed up with another patent application. The improved parking meter had a sleek, art deco look and a small, curved window that revealed expired time. He assigned his rights to Magee's Dual Parking Meter Company, which still operates as POM Inc. MaGee's design remained relatively unchanged for almost 60 years until POM introduced the electronic parking meter.

On July 16, 1935, Oklahoma City installed the first set of parking meters on one side of a downtown street. Shopping traffic increased to such an extent, that merchants on the other side of the street demanded their meters. The city reaped direct benefits as well. By the end of the month, a former mayor paid a one dollar fine for parking too long in the new five cent meter zones.

The Theremin

In the early 20th century, Russian cellist and electronic engineer Lev Sergeivitch Termen heard a peculiar sound while he assembled a radio. He discovered that simply waving his hands near the apparatus altered the frequencies of sound. This discovery inspired Termen to develop a new type of musical instrument.

Termen demonstrated his Aetherphone, the first electronic musical instrument, at a Moscow industrial fair in 1922. Vladimir Lenin learned about the strange device and invited Termen to his office in the Kremlin. After the demonstration, Lenin played the Aetherphone himself.

In 1927, Termen performed his electronic instrument at New York's Plaza Hotel. The audience enthusiastically listened to the ethereal, violin-like sounds that emanated from the Theremin, a label based on Leon Theremin, Termen's adopted name. Theremin soon licensed the Radio Corporation of America to produce his invention.

The Theremin, basically a box of radio tubes and two antennae, proved a challenging instrument to play. It lacked piano keys, strings, or other traditional reference points. A musician did not touch the Theremin, but instead controlled volume by moving the left hand around a horizontal loop antenna, and pitch by moving the right hand near an upright antenna.

Composers included the Theremin in their symphonies. Many became familiar with the Theremin's distinctive undulating tones from 1950s science fiction movies, such as *The Day the Earth Stood Still*. Television themes have incorporated Theremin-like sounds from the 1960s *Star Trek* series to the more recent *Midsomer Murders*. The Beach Boys wove the Theremin sound into the fabric of popular culture with their hit, "Good Vibrations."

Lipstick

For thousands of years, the application of paint to lips was reserved for the wealthy, courtesans and prostitutes. Around 3,500 BC, Sumerian Queen Schub-ad colored her lips with a mixture of crushed red rocks and toxic white lead. Affluent ancient Greeks also used a toxic concoction: rouge made with cinnabar, red mercury sulfide. For centuries, people applied the red cream, which was absorbed or ingested. The use of mercury-based rouge could have caused many miscarriages, stillbirths and congenital deformities.

By the 17th century, religious criticism of lip painting gained hold, especially in England. Condemnation of the practice reached its peak in Victorian England; the Queen herself publicly declared makeup "impolite." With the notable exceptions of prostitutes and stage actors, the practice of coloring lips almost disappeared from Europe by the late 19th century.

Lip painting experienced a rebirth during the 1920s in North America. Women began to color their lips in bold colors, a sign of confidence that accompanied new wealth and independence. Hollywood makeup artists Max Factor and Helena Rubinstein used their experience to market affordable cosmetics to the public. The Jazz Age also marks the release of the first color-saturated Technicolor films, an event that boosted sales of brightly-colored lipstick.

Not only the practice of lip painting changed, lip coloring technology evolved as well. In the early 20th century, sticks of lip rouge could be purchased in paper tubes, an invention of the French company, Guerlain. In 1915, American Maurice Levy introduced a lipstick encased in metal with a slide lever, push-up mechanism. Eight years later, James Bruce Mason Jr. patented the familiar case that raised the lipstick with a turn.

Safety Pin

Before the age of metal, humans fastened clothes with pins made of thorns, wood splinters, or fish bones. When metal pins first became available, they retained the flaw of earlier closures. For effective fastening, thread had to be wrapped around both sides of the pin.

Eventually, metal workers bent the pin to bring the ends together, added a spiral loop at the bend that functioned like a spring, and hammered the pin head into a groove or catch to hold the other end. Historians speculate that this fibula, an ancient form of safety pin, might have been an Aegean invention that spread to Europe around the 13th century BC.

During the Victorian era, the pin as a fabric fastening device was largely replaced by buttons, ties, toggles, and hook and eye closures. It was time for a reinvention of the safety pin.

In 1842, Thomas Woodward of New York obtained a patent for a diaper pin that had a cup at one end to hold the pointed end. The pin's two arms met at a hinge in the bend. Fabric had to be gathered between the arms to lock the pointed end in the cup. Insufficient pressure allowed the pointed end to slip out of the cup.

Walter Hunt, another New York inventor, was playing with a piece of wire when he thought of the self-sprung safety pin. In 1849, he obtained a patent for a pin made from one piece of wire, coiled into a spring in the middle. The tension of the spring forced the pin's pointed end into a clasp at the other end. Hunt's patent could have made him a rich man. But he assigned his patent rights for $400.

Sunglasses

Inuit tribes in the Arctic Circle get credit for inventing a type of sunglasses thousands of years ago. To reduce glare reflected from snow and ice, they wore pieces of whalebone with narrow horizontal slits across the center. By the 14th century, judges in China donned glasses with smoked quartz lenses. They did not wear the glasses to cut down on glare, but rather to conceal their reaction to evidence presented in court.

Colored lenses became available in 18th century Europe. James Ayscough, an English crafter of scientific instruments, produced some of the first tinted lenses in 1752. He recommended green- or blue-colored lenses to correct certain vision impairments.

In North America's early 20th century, the use of tinted glasses to avoid glare started to become popular. Silent movie stars wore colored lenses to cope with harsh studio lighting. Yet sunglasses really didn't take off until pilots started wearing them.

By the 1930s, pilots flew at increasingly higher altitudes and for longer periods of time. Sunlight glare became a serious problem. The US Air Force commissioned Bausch & Lomb to create high performance sunglasses to protect fighter pilots' eyes.

The company designed large, teardrop-shaped lenses tinted dark-green to absorb the brightest part of the spectrum. The oversized lenses sat in frames that extended downward to protect aviator's eyes when they glanced at their instrument panels. The sunglasses met with such approval that the company sold Ray-Ban aviator sunglasses to the public in 1937.

In the same year, American inventor Edwin Land founded the Polaroid Corporation. By the end of the decade, the company manufactured Polaroid sunglasses, based on Land's invention of a polarizing filter.

Since the Second World War, sunglasses have been promoted as an essential fashion accessory and as a means to protect eyes from sun damage. Either view supports the large sunglasses industry.

Transparent Tape

In 1921, Richard G. Drew joined the Minnesota Mining and Manufacturing Company, a small business that specialized in sandpaper. Four years later, Drew invented a new type of product for 3M: masking tape. An auto body shop painter tested a prototype of the tape that had adhesive only on its edges. Before the painter could apply a second color to a two-tone car, the tape peeled off. "Take this back to your stingy Scotch bosses," the painter said, "and tell them to put more adhesive on it." Drew remembered the ethnic slur.

Du Pont had recently invented water-tight cellophane. Drew thought that he could use it to make a moisture-resistant sealant for refrigerated railroad cars. In June 1929, he ordered 100 yards of cellophane and soon discovered that the material resisted adhesive as well as water. Drew and other 3M researchers formulated a mixture of rubber, oils and resin as an adhesive coating on cellophane. They also designed machinery to prevent cellophane from splitting while applying adhesive. Recalling the irate car painter, Drew named the product Scotch Tape.

In 1930, 3M marketed Scotch® Cellophane Tape to grocers, meat packers and bakers, who could use it to seal cellophane bags of food. When Du Pont devised a method of heat-sealing cellophane bags, 3M focused on the general public. Introducing a new product during the Depression seemed to be risky. Yet the tape turned out to be the perfect product for the time. People found many uses for Scotch Tape from mending toys, ripped fingernails, torn book pages, and ceiling plaster to patching cracked turkey eggs.

A rival arose in 1937 England. Colin Kininmonth and George Gray coated cellophane with a natural rubber resin, creating a product they called Sellotape®. Today, 3M shares the world market for transparent tape with its English cousin.

Automatic Dishwasher

During the late 19th century, Josephine Garis Cochran became irritated by the way that her servants' dish washing chipped and broke her heirloom china. Cochran, a Shelbyville, Illinois society hostess, decided to do something about it. She tried to find a mechanical dishwashing machine, but none were available. In 1850, Joel Houghton had earned a US patent for such a machine. But the wooden apparatus with its hand-turned wheel lacked practicality.

Cochran decided to build her own machine. She worked in her garden woodshed, twisting wire into racks to hold dishes. The racks fit on a motorized wheel in a large copper boiler. While the wheel turned, hot soapy water squirted through the bottom of the boiler and coated the dishes. To market her invention, Cochran founded the Garis-Cochran Manufacturing Company. The company manufactured two versions of dishwashers: a small foot pedal model and a large steam-driven machine that could wash and dry 200 dishes.

At the 1893 Chicago World's Fair, the Cochran dishwasher earned an award for design and durability. The company should have thrived. Instead, it only managed to survive by selling the dishwashers to hotels and restaurants. The domestic market showed disinterest. Many houses lacked hot water heaters sufficiently large to supply the dishwasher. Also, housewives at that time admitted that washing dishes was one of their more relaxing chores.

Cochran's company became part of KitchenAid. In 1949, that business introduced a home dishwasher based on the Cochrane invention. During the next decade, a change in attitude about housework, improved water-heating technology and prosperity created a demand that turned the dishwasher into standard equipment in the American home.

Monopoly® Game

It's a legendary rags-to-riches story. During the Great Depression, unemployed Pennsylvania engineer Charles B. Darrow created Monopoly and became the first millionaire inventor of a game.

Darrow drew the game design on an oil cloth in 1933. Inspired by holidays spent in Atlantic City, New Jersey, he named sections of the game after that city's streets. Charms from his wife's bracelet and bits of wood served as playing pieces.

When Darrow saw how his family and friends enjoyed the game, he thought that he had a winner. Yet Parker Brothers rejected the Monopoly game, finding over 50 design errors.

After enlisting help from a printer, Darrow produced and sold 5,000 handmade games to a Philadelphia department store. Soon, demand outstripped supply. Darrow returned to Parker Brothers. This time, they bought it. The company sold its first Monopoly game set in 1935. Since then, Monopoly has been manufactured in more than 25 languages and played by more than 450 million people worldwide.

The game has attracted a devoted following. In 1972, Atlantic City's Commissioner of Public Works proposed to change two street names that appear in Monopoly. Hundreds of game players protested the proposal. "Would you be willing to take the responsibility," Parker Brothers' president Edward P. Parker warned, "for an invasion by hordes of protesting Monopoly players, all demanding that you go directly to jail, without even the dignity of passing GO?" In a unanimous vote, the Commission retained the classic street names. Ironically, Hasbro Toys, the current producer of the game, announced in 2006 that it would drop Atlantic City references from its flagship version of Monopoly. This time, Atlantic City officials protested.

Iditarod

Hailed as the "Last Great Race on Earth," the Iditarod pits mushers and their dog teams against Alaska's subzero temperatures over 1,150 miles of forests, frozen rivers, mountain passes and tundra. During a period of 10 to 17 days, they must cover this terrain, from Anchorage to Nome on the Bering Sea coast.

In 1964, Dorothy G. Page, chair of the Wasilla-Knik Centennial, thought up the idea of a race over the Iditarod Trail. After two short trial runs in the late 1960s, the Iditarod Trail Sled Dog Race debuted in 1973. The race honors a vital part of Alaska's history.

Beginning in the 1880s, more than 30 gold rushes brought hordes of people to Alaska. In the summer, the new inhabitants could travel by steamboats on Alaska's rivers. From October to May, however, the rivers froze. In 1908, the US Army's Alaska Road Commission surveyed and constructed the Iditarod Trail for dog sled teams to transport freight from Seward to Nome. The route comprised a network of trails used by ancient native hunters, Russian explorers and the gold seekers.

Through the First World War and into the early 1920s, the Iditarod Trail served as a crucial route during Alaska's winter. In the harsh weather, dog teams provided the only means to travel between villages and to transport mail and supplies.

In 1924, aviator Carl Ben Eielson delivered Alaska's first airmail shipment. Airplanes soon dominated travel within Alaska. Yet one year later, dog teams once again showed their value when a diphtheria epidemic threatened icebound Nome. Since weather foiled plans to send diphtheria vaccine by airplane, a relay of 20 mushers and their dog teams carried the vaccine over the Iditarod Trail. In temperatures rarely warmer than 40 degrees below zero, they traveled 674 miles in about six days, saving hundreds of lives.

Steam Tractor

The Industrial Revolution brought mechanization to farms in the form of external combustion engines that produced steam power. In the 1830s, British farmers used small steam engines built on skids or wheels. Since they lacked self-propulsion, teams of horses had to pull the machines to locations where a farmer needed belt power. Burning coal or wood generated steam in the engine's boiler, and the steam drove the turning of a wheel. A belt attached to the steam engine's wheel connected to the wheel of a mechanism, such as a thresher or treadmill, which required power.

In the 1850s, North American farmers began to use the horse-drawn steam engines. Horses caught a break when manufacturers produced steam engines with a steam-powered drive train. The self-propelled machinery equipped with tractor drive steam engines eventually became known as tractors.

Steam traction engines proved far more productive than horses, but they cost a great deal more. The machines also had drawbacks. Weighing 10 to 25 tons, the heavy machines bogged down in muddy fields and could burst through a rural bridge. The boiler, which had to be cleaned frequently, needed a regular supply of clean water.

The steam tractor reached its peak of popularity between 1908 and 1915. Recognizing the deficiencies of steam power, inventors were developing a relatively light and inexpensive tractor driven by a gasoline-powered internal combustion engine. Labor shortages during the First World War accelerated the adoption of the new tractors. After the war, the more efficient and reliable gas tractor ended the steam tractor's brief reign.

The steam tractor's era has passed, but the machine has not been forgotten. Aficionados around the globe keep the steam tractor and its history alive.

A steam tractor made by the Waterloo Manufacturing Company Ltd. (Waterloo, Ontario), circa 1911. Source: Western Development Museum/Library and Archives Canada/PA-038420.

The Mystery of the *Mary Celeste*

On the afternoon of December 5, 1872, the brigantine *Dei Gratia* sailed about 400 miles from the coast of Portugal. At the helm, John Johnson spotted a vessel off the port bow – perhaps six miles away. Even at that distance, the yawing ship seemed to be in distress.

The *Dei Gratia*'s sailors watched the strange ship. For about an hour, it staggered along with only a few sails unfurled. When they passed within 300 or 400 yards, Captain Morehouse hailed her. He received no reply. The captain ordered *Dei Gratia*'s lifeboat readied; Oliver Deveau, John Johnson and John Wright would aid the mystery ship's crew.

The three men rowed to the listing ship and clambered aboard. Nobody greeted them on the creaking deck. A door slammed open and shut with the motion of the sea. The ship's neglected wheel gyrated by itself and a frayed sail hung from the foremast. Deveau, Johnson and Wright had boarded a ghost ship, the *Mary Celeste*.

For more than a century, the deserted ship spawned conjectures. Had the *Mary Celeste*'s crew mutinied? Did pirates kill everyone on board? Perhaps the tentacles of a giant octopus plucked all life from the *Mary Celeste*. A waterspout could have hit the ship or aliens might have abducted the crew. Even the Bermuda Triangle, despite its location on the other side of the ocean, has been thrown into the mix.

Much speculation focuses on *how* the ship came to be deserted. Yet this is known; everybody abandoned the *Mary Celeste*. The real mystery is *why* did they do so?

The Rise of a Ghost Ship

The *Mary Celeste* had a blighted career. Joshua Davis built the two-masted, 103-foot long vessel in the coastal village of Spencer's Island in Nova Scotia. The ship, originally named *Amazon*, launched on May 18, 1861. According to some reports, the *Amazon* first sailed to the city of Windsor to pick up a load of plaster bound for New York. But Robert McLellan, the *Amazon*'s first skipper and part-owner, developed pneumonia. The *Amazon* returned to Spencer's Island, where McLellan soon died.

In 1867, a sudden gale seized the *Amazon* from its anchorage and battered the ship against a rocky shore. The ship's owners sold the *Amazon* as a wreck to Alexander McBean. After refloating the vessel, McBean sold the ship to John Howard Beatty, who lost little time unloading it.

In November 1868, Captain Richard W. Haines bought the ship at auction for a mere $1,750. Haines invested more than $8,000 for repairs and registered the ship as the *Mary Celeste* in New York on December 31, 1868.

Captain Benjamin Spooner Briggs took command of the *Mary Celeste* in October 1872. On his first voyage, Briggs had to deliver 1,700 barrels of alcohol consigned to H. Mascarenhas and Company of Genoa, Italy. Briggs took his wife Sarah and Sophia Matilda, their two-year-old daughter.

On November 5, the ship with her crew of seven departed from pier 50 on the East River. Rough weather forced the captain to seek shelter for two days before setting out again. It was an inauspicious beginning.

Around November 24, the *Mary Celeste* battled a gale and heavy rain as it sailed near the islands of the Azores group. Meteorological records indicate that, on the forenoon of the 25th, a calm prevailed, but that this soon turned into gale force winds.

When the *Dei Gratia* sailors boarded the *Mary Celeste* on December 5, they found an eerie sight: The captain's cabin contained an unmade bed that bore the impression of a sleeping child. They could find no signs of fire damage or violence. The pantry held about six months' worth of food and plenty of water. The ship appeared seaworthy and abundantly provisioned.

Through the open fore-hatch, Deveau noticed barrels labeled "alcohol" in the cargo hold. The lazaret hatch, which led to a small space below the main deck, also stood open. The hold carried three and one-half feet of water. Deveau later concluded that the water had entered through the open hatches after the crew had left and nobody manned the pumps.

The sailors noticed something missing: a small yawl that served as a lifeboat. It seemed that the captain, his family and the crew had quickly vacated their ship.

A skeleton crew from the *Dei Gratia* sailed the *Mary Celeste* to Gibraltar. Here, they claimed salvage rights.

Rumors erupted. On December 21, the *Shipping and Commercial List* reported the rescue of the derelict *Mary Celeste*. "The inference is that there has been foul play somewhere, and that alcohol is at the bottom of it."

On January 30, 1873, the *Gibraltar Chronicle* noted the discovery of a blood-smeared sword on the *Mary Celeste*. This news fueled imaginations.

"The conclusion seems inevitable that the crew had mutinied," reported Ohio's *Richmond Gazette* of March 20, 1873, "that after a fierce struggle they had seized the brig, murdered the captain, his wife, and probably the subordinate officers, and taken to the boats."

One man in particular became convinced that *Mary Celeste*'s crew had become brutal cutthroats: Frederick Solly Flood, the Attorney-General for Gibraltar. During salvage hearings in the Vice Admiralty Court, Flood urged that the crew, drunk from the ship's cargo, had murdered the Chief Mate, the Captain and his family, and then fled to another vessel. Flood pointed to barrels of alcohol that stood empty in the hold.

Seeking to bolster his theory with scientific evidence, Flood obtained the services of Dr. J. Patron. Yet Patron could not find signs of violence and concluded that blood had not stained the sword.

The US Consul, Horatio Sprague, requested Captain R.W. Shufeldt to inspect the ship. Shufeldt found no evidence of violence and nothing to indicate a mutiny.

At the end of the hearing, the judge awarded the *Dei Gratia* salvagers 1,700 minus the cost of bloodstain analysis. The salvage had not been worth their effort.

Once again, the *Mary Celeste* changed hands. The remaining part-owners sold their shares to James H. Winchester, who sold a quarter share to J.Q. Pratt, the new skipper. In February 1874, they sold the ship to a consortium. Six years later, Wesley A. Grove bought the *Mary Celeste*. In December 1884, Captain Gilman C. Parker set sail for Port-au-Prince, the capital of Haiti. He ran the *Mary Celeste* into the rocks of Rochelois Bank in the Gulf of Gonave. The ship ended her life to further an insurance swindle.

The insurance companies' investigators learned that the *Mary Celeste*'s crew had filled the holds with dog collars, cheap rubber overshoes, waste beer, rotten fish and other garbage. The cargo had been falsely insured for significantly more than its value.

A trial during the summer of 1885 ended with a hung jury. Captain Parker died while waiting for his new trial, and US Attorney George P. Sanger decided that the government would no longer pursue the matter.

A Spirited Theory

In his book, *Ghost Ship*, Brian Hicks proposes that the clue for the abandonment of the *Mary Celeste* lies in its cargo. The American consul in Genoa reported nine barrels of alcohol – 450 gallons – had leaked during the voyage. Frederick Solly Flood had suggested that the barrels contained distilled spirits, but Oliver Deveau had testified that the *Mary Celeste* carried no spirits.

The cargo was probably industrial alcohol, such as methanol. Fumes from more than 400 gallons of methanol would have made the crew ill and forced them to find fresh air.

During the salvage hearing, the *Dei Gratia* sailors described *Mary Celeste*'s open hatches, doors, skylight and windows. The missing crew had tried to air out the ship. But weather reports show the sea becalmed. In the dead air, they had one course of action: Leave the *Mary Celeste* and wait for the fumes to dissipate.

The *Dei Gratia* sailors testified that the halyard – a rope used to hoist a sail – hung over the ship's side with a torn end. This suggests, says Hicks, that Briggs ordered his men to fasten the peak halyard of the mainsail to the front of the launch. In this way, the crew could haul the small boat back to the ship. Using the peak halyard, one of the longest lines on board, the crew could have drifted 200 or more feet from the ship.

Unfortunately, the calm weather dramatically transformed. A gale arrived. Before evacuating the ship, the sailors had been so rushed or so ill from the fumes, that they left sails unfurled and did not tie off the wheel.

The wind caught the ship's sails; waves controlled the ship's course. As the vessel picked up speed, the crew struggled against a violent sea to haul the overcrowded craft back to the *Mary Celeste*. But their worn lifeline snapped.

The fate of the crew may never be known for certain. As Captain Shufeldt said in his report, it's a "sad and silent mystery of the sea."

Further Reading

The Discovery of Tutankhamun's Tomb

Carter, Howard and A.C. Mace. *The Discovery of the Tomb of Tutankhamen* (Dover Publications, 1977).

Hoving, Thomas. *Tutankhamun: The Untold Story* (Simon and Schuster, 1978).

Smith, G. Elliot. *Tutankhamen and the Discovery of His Tomb* (George Routledge & Sons, Ltd., 1923).

The Timeless Appeal of Clocks

Andrewes, William J.H. "A Chronicle of Timekeeping," *Scientific American* (September 2002).

Barnett, Jo Ellen. *Time's Pendulum* (Harcourt, Inc., 1998).

Landes, David S. *Revolution in Time* (Belknap Press, 2000).

Viredaz, Michel. "An Introduction to Electric Clocks," *NAWCC Bulletin* (April 2003).

Plastics

Lokensgard, Erik. *Industrial Plastics, Fourth Edition* (Delmar Learning, 2004).

Meikle, Jeffrey I. *American Plastic* (Rutgers University Press, 1995).

Mossman, Susan (ed.). *Early Plastics* (Leicester University Press, 1997).

Darwin Aboard the *HMS Beagle*

Browne, Janet. *Charles Darwin: Voyaging* (Princeton University Press, 1995).
The Complete Works of Charles Darwin on the *Darwin Online Website.*
Darwin, Charles. *Voyage of the Beagle* (Penguin Books, 1989).
Darwin Correspondence Project Website (2013).

The Shocking History of Electricity

Jonnes, Jill. *Empires of Light: Edison, Tesla, Westinghouse, and the Race to Electrify the World* (Random House, Inc., 2003).
Meyer, Herbert W. *A History of Electricity and Magnetism* (Burndy Library, 1972).
Schewe, Phillip F. *The Grid: A Journey Through the Heart of Our Electrified World* (Joseph Henry Press, 2007).

The Smith Butchering Machine

"Edmund A. Smith – In Memorium," *Pacific Fisherman* 7(6):19, 21 (June 1909).
"'Iron Chink' a Notable Factor in the Advancement of the Salmon Industry," *Pacific Fisherman*, 112-113 (January 1927).
"'Iron Chink' a Winner Among Fisheries of Two Coasts," *Pacific Fisherman* 4(5):13 (May 1906).
O'Bannon, Patrick W. "Technological Change in the Pacific Coast Canned Salmon Industry, 1900-1925: A Case Study," *Agricultural History* 56(1):151-166 (1982).
"The Smith Fish Cleaning Machine," *Pacific Fisherman Annual* 4:45-47 (June 1906).
Wilson, Margaret and Jeffery L. MacDonald. "The Impact of the 'Iron Chink' on the Chinese Salmon Cannery Workers of Puget Sound," *The Annals of the Chinese Historical Society of the Pacific Northwest*, 79-89 (1984).

Diabetes Treatment Comes of Age

Bliss, Michael. *The Discovery of Insulin* (University of Toronto Press, 1982).

"The Discovery and Early Development of Insulin," *University of Toronto Library Website* (2003).

Revealing Cryptic History

Samuel, Dorn, V. "Code Breaking in Law Enforcement: A 400-Year History," *Forensic Science Communications* (April 2006).

Singh, Simon. *The Code Book* (Random House, Inc., 1999).

Wrixon, Fred B. *Codes, Ciphers, & Other Cryptic & Clandestine Communication* (Black Dog & Leventhal, 1998).

The Mystery of the *Mary Celeste*

Begg, Paul. *Mary Celeste* (Pearson Education Ltd., 2005).

Hicks, Brian. *Ghost Ship* (Ballantine Books, 2004).

Lightning Source UK Ltd.
Milton Keynes UK
UKHW01f1935300418

321899UK00048B/1679/P